Practice in BIOLOGY

Progressive questions for AS and A level

Hodder & Stoughton

A MEMBER OF THE HODDER HEADLINE GROUP

Orders: please contact Bookpoint Ltd, 130 Milton Park, Abingdon, Oxon OX14 4SB.
Telephone: (44) 01235 827720, Fax: (44) 01235 400454. Lines are open from 9.00–6.00, Monday to Saturday, with a 24-hour message answering service. You can also order through our website at www.hodderheadline.co.uk

British Library Cataloguing in Publication Data
A catalogue record for this title is available from The British Library

ISBN 0 340 77267 0

First published 2000

Impression number	10 9 8 7 6 5
Year	2006 2005 2004 2003

Cover design by Blue Pig Design Co.
Illustrated by Tom Cross.
Typeset by Wearset, Boldon, Tyne & Wear.
Printed in Great Britain for Hodder & Stoughton Educational, a division of Hodder Headline Ltd, 338 Euston Road, London NW1 3BH by JW Arrowsmith, Bristol.

Photo acknowledgments
The publishers would like to thank the following individuals, institutions and companies for permission to reproduce photographs in this book. Every effort has been made to trace ownership of copyright. The publishers would be happy to make arrangements with any copyright holder whom it has not been possible to contact:

Biophoto Associates (4, 5, 28, 29, 32, 33, 34, 84)
Science Photo Library (3)

Contents

Specification matching grid

AS topics

Chapter	*Topic*	*AQA Spec. A*	*AQA Spec. B*	*Edexcel*	*OCR*
1	Cells	Module 1	Module 1	Unit 1	Module A
1	Tissues	Module 1	Module 1	Unit 1	Module A
2	Cell membrane	Module 1	Module 1	Unit 1	Module A
2	Cell transport	Module 1	Module 1	Unit 1	Module A
3	Molecules	Module 1	Module 1	Unit 1	Module A
4	Enzymes	Module 1	Module 1	Unit 1	Module A
4	Enzymes – in industry/biotechnology	Module 2		Unit 1	
5	Nutrition and digestion			Unit 2B	
5	Nutrition – saprobionts		Module 1	Unit 3	
5	Nutrition in parasites			Unit 3	
6	Gas exchange – basic principles	Module 1	Module 1	Unit 2B	Module A
6	Gas exchange – humans	Module 1	Module 1	Unit 2B	Module A/B/C1
6	Gas exchange – fish				Module 1
6	Gas exchange – insects				
6	Gas exchange – plants		Module 1	Unit 2B	Module A
7	Transport – heart/blood	Module 1	Module 3	Unit 2B	Module B/C1
7	Transport – plants		Module 3	Unit 2B	Module C1
7	Xerophytic adaptation	Module 2		Unit 2B	
8	Nucleic acids – DNA and RNA	Module 2	Module 2	Unit 1	Module A
8	Protein synthesis	Module 2		Unit 1	Module A
8	Genes and genetic code/mutation	Module 2	Module 2	Unit 1	Module A
8	Gene technology	Module 2	Module 2		Module A
9	Cell cycle/mitosis	Module 2	Module 2	Unit 1	Module A
10	Reproduction – mammal	Module 2	Module 2	Unit 2B	
10	Reproduction – flowering plants			Unit 2B	
16	Ecosystems – basic principles			Unit 3	Module A
16	Ecology – diversity and succession				
16	Ecology – energy flow			Unit 3	Module A
16	Ecology – biological pyramids			Unit 3	
16	Carbon cycle			Unit 3	
16	Nitrogen cycle			Unit 3	Module A
16	Fertilisers and pesticides	Module 2		Unit 3	
17	Energy resources			Unit 3	
17	Deforestation and desertification			Unit 3	
17	Air pollution			Unit 3	
17	Water pollution			Unit 3	
21	Data handling	All	All	All	All

A2

Chapter	Topic	AQA Spec. A	AQA Spec. B	Edexcel	OCR
1	Cells – prokaryotes		Module 6Q		
4	Enzymes – industrial		Module 6Q		Module E4
5	Digestion	Module 6			Module E5
6	Gas exchange – fish	Module 6			
6	Gas exchange – insects	Module 6			
6	Gas exchange – flowering plants	Module 6			
7	Transport – blood and respiratory gases	Module 6			
7	Plant transport/xerophytes	Module 6			
8	Gene technology			Unit 5B	Module E2
9	Meiosis	Module 5	Module 4	Unit 5B	Module D
10	Reproduction – human		Module 6R		Module E1
10	Reproduction – flowering plants				Module E1
11	Genetics	Module 5	Module 4	Unit 5B	Module D
12	Classification	Module 5	Module 4	Unit 5B	Module D
13	Variation	Module 5	Module 4	Unit 5B	Module D/E2
13	Selection	Module 5	Module 4	Unit 5B	Module D
13	Selection and Hardy–Weinberg	Module 5			
13	Speciation/evolution	Module 5	Module 4	Unit 5B	Module D
14	Respiration	Module 5	Module 4	Unit 4	Module D
15	Photosynthesis	Module 5	Module 4	Unit 5B	Module D
15	Photosynthesis – limiting factors		Module 6P	Unit 5B	Module D
16	Ecosystems – basic principles	Module 5	Module 6P		Module E3
16	Ecology – diversity and succession	Module 5	Module 6P	Unit 5B	Module D
16	Ecology – ecological pyramids and energy flow	Module 5			Module 5a
16	Ecology – populations	Module 5	Module 5a	Unit 5B	Module D
16	Ecology – techniques		Module 5a	Unit 5B	Module D/E3
16	Carbon cycle	Module 5	Module 5a		
16	Nitrogen cycle	Module 5	Module 5a		
17	Deforestation	Module 5			
17	Air pollution		Module 6P		Module E3
17	Water pollution		Module 6P		Module E3
17	Ecology and farming/pesticides/ biological control		Module 5a	Unit 5B Module 6P	Module D/E3
17	Fisheries		Module 6P		Module E3
17	Conservation	Module 5	Module 6P	Unit 5B	Module E3
18	Homeostasis – principles	Module 6	Module 4	Unit 4	Module D
18	Homeostasis – temperature and glucose levels	Module 6		Unit 4	Module 4
18	Homeostasis – liver function	Module 6		Unit 5B	Module E5
18	Excretion – fish/birds	Module 6			
18	Excretion – kidney in mammals	Module 6	Module 4	Unit 4	Module D
18	Osmoregulation	Module 6	Module 4	Unit 4	
19	Nervous system – neurones, action potentials, nerve impulse, synapse	Module 6	Module 4	Unit 4	Module D
19	Nervous system – receptors	Module 6	Module 4	Unit 4	Module D/E5
19	Nervous system – reflexes and behaviour	Module 6	Module 4	Unit 4	Module E5
19	Chemical control – hormones	Module 6	Module 4	Unit 4	Module D
19	Chemical control – plants		Module 4	Unit 5B	
20	Synoptic	Module 8	Unit 6	Unit 5/6	Module F1
21	Data handling	All	All	All	All

Using this book

Each chapter starts with a section titled 'You should already know that …'. This is a short overview of the knowledge that you should have gained during GCSE and bring with you to AS and A level. The Key Facts is a list of important points that summarise a particular topic. You should bear in mind that the points listed are not necessarily specific to the specification that you are studying.

The questions are intended to be used during the course and not just at the end of the course. The first questions in each section are really review questions to remind you of the knowledge that you should have gained during GCSE. The questions move through the subject area, getting more difficult. There are some longer free prose and essay-style questions. There are some questions to test your skills in experimental biology. Synoptic questions, which cover a number of different topic areas, are provided at the end of the book.

The last chapter covers data handling. The biology specifications list the mathematical skills that you will need in order to answer questions in the examinations. The data handing chapter covers all of these requirements and there are a number of worked examples to show you how to carry out simple statistical tests.

1 Cells

You should know that:

- living organisms are made up of units called cells
- animal and plant cells have a cell membrane, cytoplasm and a nucleus; plants have, in addition, a cellulose cell wall, vacuole and in some cells, chloroplasts
- the structure of a cell is related to its function.

KEY FACTS

- The cell is the basic unit of life.
- Cells are surrounded by a partially permeable membrane which encloses the cytoplasm. Within the cytoplasm are organelles.
- Cell organelles have specific functions. There is division of labour leading to greater efficiency.
- The nucleus is the largest organelle, surrounded by a double nuclear membrane, perforated by pores. It contains nucleoplasm and a nucleolus.
- The endoplasmic reticulum (ER) is a 3D network of cavities, bounded by membranes. Rough ER is concerned with the isolation and transport of proteins. Smooth ER is involved with the synthesis and transport of lipids and steroids.
- The ribosome is the site of protein synthesis.
- The Golgi apparatus consists of stacks of parallel, flattened membrane-bound sacs. It processes proteins.
- Mitochondria are rod-shaped structures with a double membrane. They are involved with aerobic respiration.
- Lysosomes are tiny membrane-bound vesicles containing hydrolytic enzymes for the breakdown of unwanted organelles or for cell self destruction.
- Centrioles are only found in animal cells where they help to form the spindle during nuclear division.
- Plant cells also contain chloroplasts, a large central vacuole and are surrounded by a cell wall.
- Much of the cell structure is not visible under the light microscope because of the limit of resolution of the wavelength of light. Cell organelles are visible using an electron microscope as electrons have a shorter wavelength.
- There are two types of cell, prokaryotic cells which lack a nucleus and eukaryotic cells that have a nucleus. Prokaryotes are bacteria. Eukaryotes include animals, plants and fungi.
- Prokaryotic cells have no nuclear membrane, no mitochondria and endoplasmic reticulum, smaller ribosomes and few other organelles.

- Cells undergo differentiation to become adapted for specialised jobs.
- Tissues are collections of cells which have a particular role. An organ is composed of more than one tissue and is a structural and function unit.
- Cells range in size from 5–20 μm. The following units are used in the study of cells:

  1 mm (millimetre) = 1/1000 of a metre = 0.001 m = 1×10^{-3} m
  1 μm (micrometre) = 1/1 000 000 of a metre = 0.000 001 m = 1×10^{-6} m
  1 nm (nanometer) = 1/1 000 000 000 of a metre = 0.000 000 001 m = 1×10^{-9} m

Questions for you to try

1.1 What is a cell? **[2]**

1.2 If you looked at an animal and a plant cell under a light microscope what differences would you see? **[4]**

1.3 Figure 1.1 shows the structures which would be visible in a plant cell examined under an electron microscope.

(a) Identify the parts labelled in this plant cell. **[13]**

(b) State one function for B, C, G, K, N. **[5]**

(c) What is part J made of? **[1]**

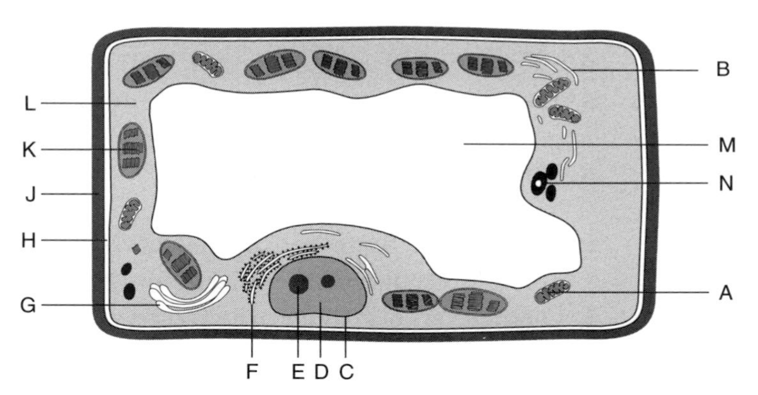

Figure 1.1

1.4 What is a tissue? Give **two** examples of tissue you would find in a human body. **[3]**

1.5 How does an organ differ from a tissue? **[1]**

1.6 The photograph in Figure 1.2 shows an organelle of a cell as seen under an electron microscope.

(a) Name this organelle. **[1]**

(b) What is the function of this organelle? **[1]**

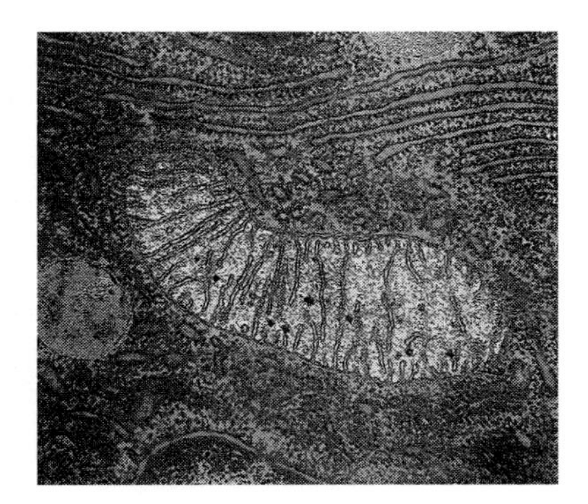

Figure 1.2

(c) This organelle has been magnified by ×12 000. Calculate the actual width of this organelle across the widest point. **[2]**

1.7 What is the difference between rough and smooth endoplasmic reticulum? **[2]**

1.8 Describe the path of a newly synthesised protein from a ribosome to the cell surface membrane. **[4]**

1.9 What is the role of the cytoskeleton in a cell? **[2]**

1.10 How would you set up a light microscope to view a slide under high magnification? **[4]**

1.11 Give **two** advantages of the electron microscope over a light microscope **[2]**

1.12 What is the difference between magnification and resolution? **[3]**

1.13 Copy and complete the following table with ✓ and ✗ to indicate whether the statements are true or false. **[5]**

	Prokaryote	*Eukaryotic animal cell*	*Eukaryotic plant cell*
cell wall is made of polysaccharide			
DNA enclosed			
mitochondria present			
cell membrane present			
ER present			

1.14 What features of a liver cell tell you that it is a metabolically active cell? **[2]**

1.15 The technique of cell fractionation is used to separate the different organelles within a sample of liver tissue and is summarised in Figure 1.3. First, the liver tissue is homogenised to break up the cells. The cells are suspended in ice cold isotonic solution and centrifuged at a low speed. All the sediment collects at the

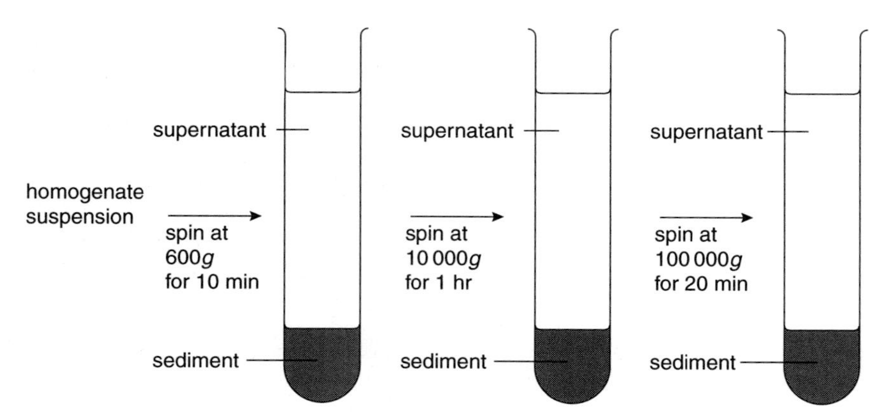

Figure 1.3

bottom of the centrifuge tube. The supernatant is poured into another tube and centrifuged at a higher speed. The process is repeated several times, and each time the tube is spun at a higher speed.

(a) Why is the liver tissue suspended in an isotonic solution? **[1]**

(b) Why is the isotonic solution ice cold? **[1]**

(c) Why is the nucleus the first cell component to separate out at the lowest speed? **[1]**

(d) Which organelle would be the last to separate out? **[1]**

(e) How could you check that the cell organelles have been separated? **[1]**

1.16 The photo in Figure 1.4 shows a group of epithelial cells from the small intestine. How are these cells specialised to carry out their function? **[2]**

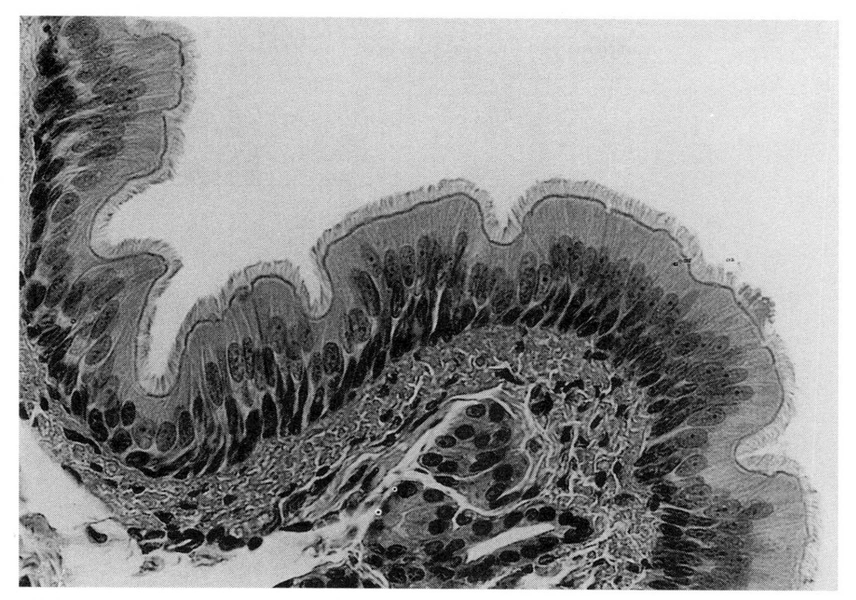

Figure 1.4

1.17 The photo in Figure 1.5 shows palisade cells in a plant leaf.

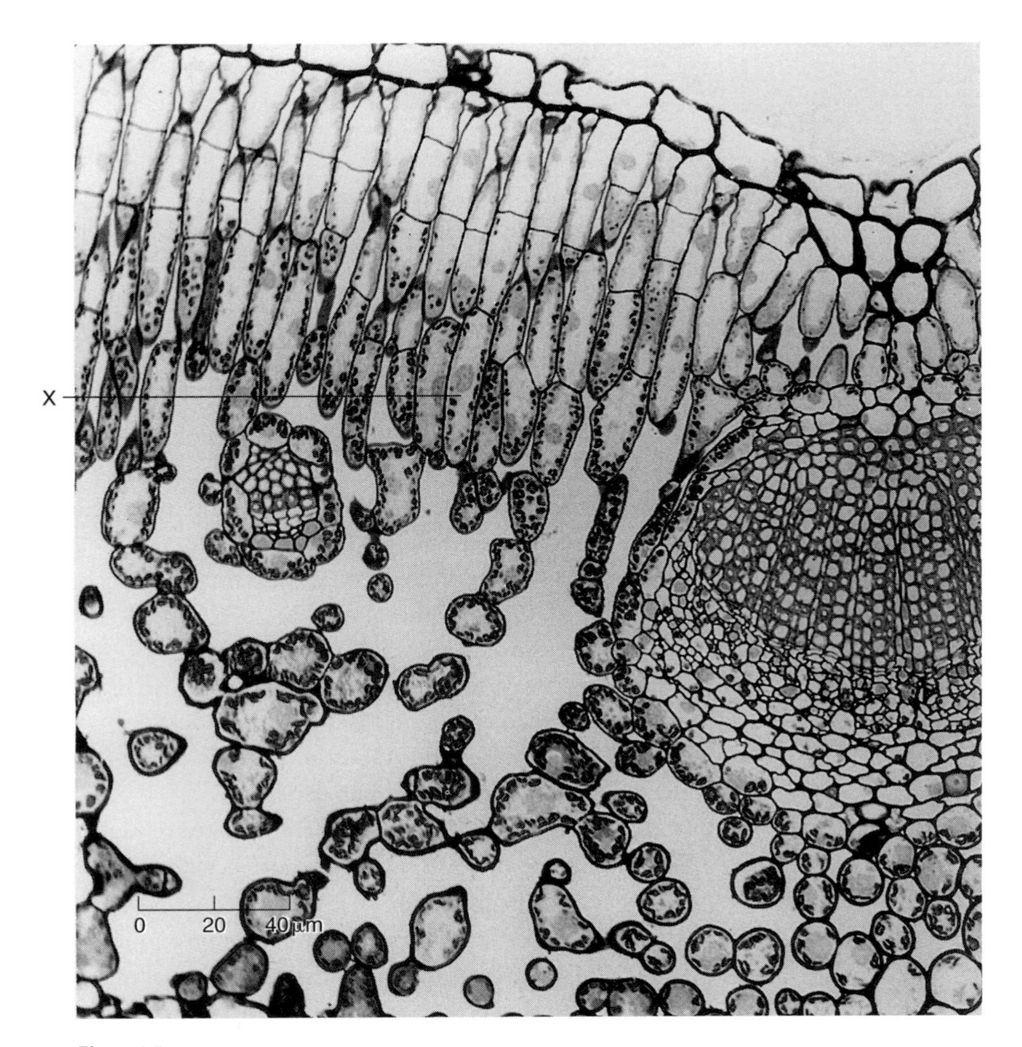

Figure 1.5

(a) Make a detailed drawing of the cell marked X. **[4]**

(b) Label your drawing and indicate on your labels the ways in which this cell is specialised for its role. **[4]**

1.18 Describe how you would make a temporary microscope mount of simple pavement epithelial cells taken from the lining of your mouth. In your answer, refer to any staining technique that you could use. **[4]**

How would you estimate the size of these cells? **[3]**

2 Membranes

You should know that:

- cells are surrounded by a membrane
- diffusion is the movement of molecules from an area of high concentration to one of lower concentration
- osmosis is the movement of water from an area of high concentration to one of lower concentration, through a partially permeable membrane.

KEY FACTS

- The cell surface membrane is partially permeable and controls the entry and exit of substances to the cell.
- The membrane is made of phospholipids, proteins, glycoproteins and glycolipids and cholesterol.
- The structure is based on the fluid mosaic model.
- Within the cell, there are membrane-bound organelles. This allows compartmentation.
- Metabolic pathways take place on membranes.
- Membranes have receptor and recognition sites.
- Diffusion is the movement of ions or molecules from a region of higher concentration to one of lower concentration.
- Osmosis is the movement of water from an area of higher, or less negative, water potential to one of a lower, or more negative, water potential, through a partially permeable membrane.
- Water potential is the potential for water to move out of a solution by osmosis. Pure water has the highest water potential of 0. All solutions have a water potential that is lower than water, therefore all solutions have a water potential that is a negative value.
- Facilitated diffusion is the rapid passage of polar molecules, such as glucose and amino acid, via transport proteins in the membrane without any energy expenditure.
- Some ions are moved across a membrane against the concentration gradient by active transport, using carrier proteins and ATP.
- Large amounts of materials are moved into or out of the cell by endo- and exocytosis.

Questions for you to try

2.1 Explain the term diffusion. **[2]**

2.2 Give **three** examples of diffusion which occur in living organisms. **[3]**

2.3 What is osmosis? **[2]**

2.4 What does partially permeable mean? **[2]**

2.5 Give **one** example of osmosis in a plant. **[1]**

2.6 List **four** ways by which materials can enter a cell. **[4]**

2.7 Figure 2.1 shows the structure of a cell surface membrane.
Name the **three** structures labelled on this diagram of a membrane. **[3]**
What is the approximate width of this membrane? **[1]**

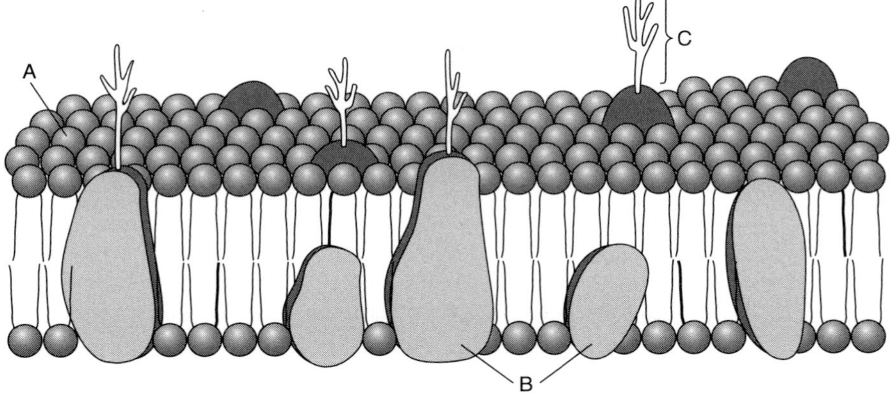

Figure 2.1

2.8 What do the words hydrophilic and hydrophobic mean? **[2]**

2.9 What is the difference between a polar and a non polar substance? **[1]**

2.10 Explain how the properties of phospholipids are important in the formation of membranes. **[2]**

2.11 There are many types of protein in a membrane. Describe the role of two proteins. **[2]**

2.12 State **two** roles of cholesterol in the membrane. **[2]**

2.13 Figure 2.2 shows a beaker containing two solutions separated by a membrane.
(a) Which solution is more concentrated? **[1]**
(b) Which solution has the higher water potential? **[1]**
(c) In which direction will the water molecules move by osmosis? Why? **[3]**

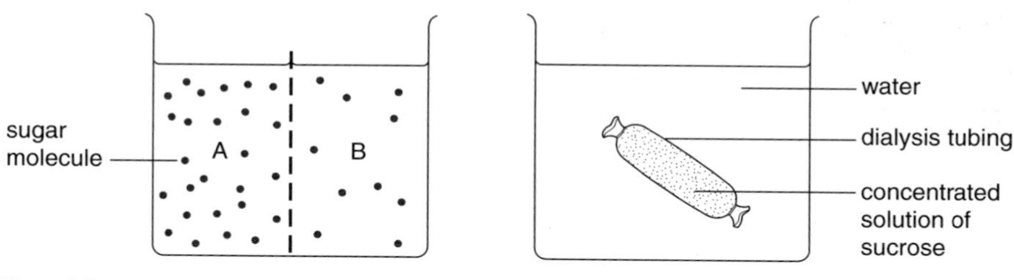

Figure 2.2

Figure 2.3

2.14 Describe how the following factors affect the rate of diffusion into a cell:

(a) concentration gradient, **[1]**

(b) molecular size. **[1]**

2.15 A piece of dialysis tubing was filled with concentrated sucrose solution and tied securely at each end. The tube was immersed in a beaker of water and left for several hours. Figure 2.3 shows the appearance of the tubing at the beginning of the experiment.

(a) What is dialysis tubing supposed to mimic? **[1]**

(b) Draw the appearance of the tubing at the end of the experiment. **[2]**

(c) Explain the changes that you have shown on your diagram. **[2]**

2.16 Distinguish between facilitated diffusion and active transport, giving an example of each. **[4]**

2.17 Figure 2.4 shows the appearance of a plant cell under the microscope. The cell membrane is pulling away from the cell wall. This is called plasmolysis.

(a) What has caused the cell to become plasmolysed? **[2]**

(b) How can this condition be corrected? **[2]**

(c) Why don't animal cells suffer from plasmolysis? **[2]**

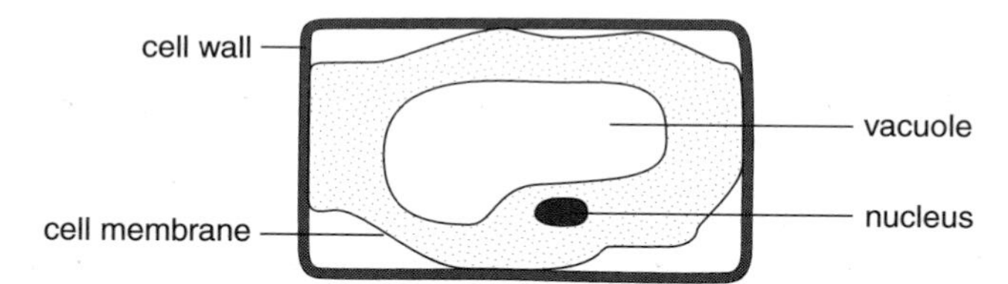

Figure 2.4

2.18 Describe the role of carrier proteins in active transport. **[2]**

2.19 The *Amoeba* is a single-celled organism that lives in water. Describe how it engulfs particles of food by endocytosis. **[2]**

3 Biological molecules

You should already know that:

- carbohydrates, such as sugar and starch, contain carbon, hydrogen and oxygen and are energy sources
- lipids, such as fats and oils, contain carbon, hydrogen and oxygen and are good energy sources
- proteins contain amino acids; proteins are used for growth and repair
- starch can be identified by the iodine test, sugars using Benedict's solution and protein using the biuret test.

KEY FACTS

- Large biological molecules are made up of many smaller units bonded together.
- Polymers are made from similar or identical sub units called monomers.
- Polymers can be built up from monomers by condensation reactions. Condensation reactions form bonds with the elimination of a molecule of water. Polymers can be broken down into monomers by hydrolysis reactions in which the links are broken by the addition of a molecule of water.
- Carbohydrates have the general formula $C_x(H_2O)_y$. Monosaccharides condense to form disaccharides and polysaccharides. Glycosidic bonds link the units.
- Monsaccharides include glucose and fructose. Disaccharides include maltose, sucrose and polysaccharides include amylose (present in starch), cellulose and glycogen.
- Glycerol and fatty acids condense to form lipids, bonded together by ester bonds. In phospholipids, one of the fatty acids is replaced by a phosphate group.
- Carbohydrates and lipids are energy sources.
- Proteins are made up of polypeptide chains. The chains consists of monomers called amino acids held together by peptide bonds. The structure of a polypeptide can be described as primary, secondary and tertiary.
- Hydrogen, ionic and disulphide bonds hold the protein molecule in shape.
- Any change to the structure of a protein can affect its activity.
- Benedict's solution tests for the presence of reducing and non-reducing sugars, iodine in potassium iodide solution tests for starch, the biuret test is used for proteins and the emulsion test detects the presence of lipids.
- Water makes up much of the cell and forms an environment in which many organisms live. Its properties make it essential for life.

Questions for you to try

3.1 Which **three** elements are found in carbohydrates, lipids and proteins? **[1]**

3.2 Give **two** examples of how carbohydrates are used in the body. **[2]**

3.3 Name **three** foods which are rich in proteins? **[1]**

3.4 State **three** functions of proteins in the body. **[3]**

3.5 What is the difference between a fat and an oil? **[1]**

3.6 What is a polymer? **[2]**

3.7 The general formula of a monosaccharide is $(CH_2O)_n$ where n is any number between 3 and 9. What would be the formula of a pentose sugar where n is 5? **[1]**

3.8 Which element is found in proteins, but not in carbohydrates? **[1]**

3.9 The results of food tests on a unknown sample are shown below. Copy and complete the table to show the conclusions which could be drawn from these tests. **[4]**

Food test	*Result*	*Conclusion*
sample mixed with iodine in potassium iodide	blue-black colour	
sample boiled with Benedict's solution	blue colour	
sample treated with dilute acid, neutralised and then tested with Benedict's	red precipitate	
sample tested using biuret solution	blue ring at surface and on shaking lilac-purple solution	

3.10 Fill in the gaps in the passage below with the most appropriate word or words:

Sugars, starches and cellulose are all examples of carbohydrates. __________ are white, crystalline and sweet tasting solids which __________ in water. They can be classified according to the number of __________ atoms present in the molecule. Sucrose and maltose are formed when two monosaccharides join together in a __________ reaction. The bond that forms between them is called a __________ bond. Sucrose is formed when a molecule of __________ bonds with a molecule of __________. Maltose is formed from two molecules of __________. Disaccharides can be converted back to monosaccharides in a __________ reaction. **[9]**

3.11 Distinguish between:

(a) alpha (α) glucose and beta (β) glucose, **[2]**

(b) glycogen and cellulose, **[2]**

(c) amylopectin and amylose. **[2]**

3.12 Why is an amino acid described as being amphoteric? **[2]**

3.13 **(a)** Draw a diagram to show how two amino acids link together to form a dipeptide. **[3]**

(b) Name the bond that links amino acids together in a polypeptide chain. **[1]**

3.14 Describe the structure of a fibrous protein such as collagen or keratin. **[3]**
How does its structure help in its function? **[2]**

3.15 **(a)** The protein, haemoglobin, has a globular structure. What does this mean? **[2]**

(b) What type of bonds hold a globular protein together? **[2]**

(c) How is the structure of a globular protein linked to its function? **[2]**

(d) What is the role of iron in haemoglobin? **[2]**

(e) What effect does high temperature have on a globular protein? **[2]**

3.16 Figure 3.1 shows the structure of a triglyceride.

(a) What are the constituents of this macromolecule? **[2]**

(b) What type of bond links the units together? **[1]**

(c) What is the difference between this molecule and a phospholipid? **[1]**

(d) State **two** functions of lipids in living organisms. **[2]**

```
        H
        |    O
        |     \\
H — C — O — C — (CH2)nCH3
        |    O
        |     \\
H — C — O — C — (CH2)nCH3
        |    O
        |     \\
H — C — O — C — (CH2)nCH3
        |
        H
```

Figure 3.1

3.17 You are provided with three solutions containing differing solutions of unknown concentrations of glucose and one solution of known concentration. Using Benedict's solution, describe how you would carry out a semi-quantitative estimate of glucose concentration. **[6]**

3.18 A solution of egg albumen was hydrolysed using the enzyme trypsin. The peptide bonds were broken and then the amino acids were separated using paper chromatography. The resulting chromatogram is shown in Figure 3.2.

(R_f = distance moved by spot ÷ distance moved by solvent)

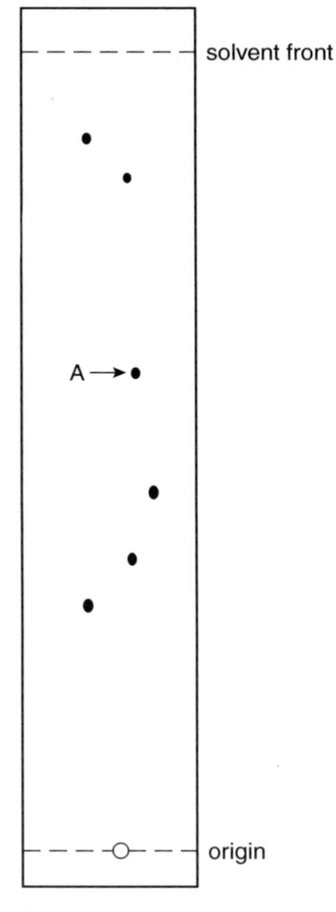

Figure 3.2

Amino acid	*R_f value*
lysine	0.14
glycine	0.26
alanine	0.38
valine	0.60
leucine	0.73

(a) Using the information provided in the table, work out the identity of the amino acid marked A on the chromatogram. **[2]**

(b) Egg albumen contains 15 different amino acids. Why didn't all 15 appear on the chromatogram? **[2]**

(c) Explain why the amino acids moved different distances along the chromatogram. **[2]**

3.19 Write a short account of the properties of water that make it essential for life. **[8]**

4 Enzymes

You should know that:

- enzymes are biological catalysts that speed up the rate of chemical reactions
- enzymes are proteins
- each enzyme has its own special shape
- each enzyme will only catalyse a particular reaction
- enzymes are affected by temperature and pH.

KEY FACTS

- Enzymes are globular proteins with a tertiary structure.
- Enzymes are catalysts that increase the rate of reaction without taking part in the reaction itself.
- Small amounts of enzyme are required as the enzyme can be used over and over again.
- Enzymes are specific, only catalysing one type of reaction.
- Enzymes have an active site with a specific 3D shape into which only one substrate can fit (the lock and key theory).
- Induced fit theory suggests that the shape of the active site changes when the correct substrate attaches.
- The substrate fits the active site to form an enzyme–substrate complex.
- Enzymes reduce the activation energy of a reaction.
- Enzymes have an optimum temperature and pH at which they work best.
- The concentration of the substrate and the enzyme affect the rate of reaction.
- Enzymes are affected by inhibitors which can be reversible or non-reversible, competitive or non-competitive (non-site directed).
- Many enzymes require cofactors to work properly.
- Industrial enzymes are used in processes which require thermostability. They are often immobilised to give more control and stability.

Questions for you to try

4.1 What is the meaning of the word 'catalyst'? **[2]**

4.2 Describe the tertiary structure of a protein. **[3]**

4.3 Why are there hundreds of different enzymes in a cell? **[1]**

4.4 Describe the active site of an enzyme. **[2]**

4.5 What is meant by the term 'lock and key'? **[2]**

4.6 Describe what is meant by the term 'activation energy'. **[1]**

4.7 How do enzymes reduce the activation energy of a reaction? **[2]**

4.8 Explain why enzymes work faster at higher temperatures. **[3]**

4.9 Describe what happens to the enzyme structure if the temperature is raised well above the optimum temperature. **[3]**

4.10 Draw a graph to show how an enzyme's rate of reaction increases with temperature. **[3]**

4.11 What does the term Q_{10} mean with respect to enzyme reactions? **[1]**

4.12 How are enzymes affected by pH? **[3]**

4.13 Why do different enzymes have a different optimum pH? **[2]**

4.14 What is the difference between a reversible and an irreversible enzyme inhibitor? **[4]**

4.15 Explain the term cofactor and give **two** examples of cofactors. **[4]**

4.16 Biological washing powders contain several different enzymes. Why should these powders be used at low washing temperatures? **[2]**

4.17 Some bacteria and algae can survive in the boiling waters of hot springs. Enzymes from these organisms are used in industrial processes. Why are these enzymes useful? **[2]**

4.18 The following set of data show the effect of temperature on the completion time of an enzyme reaction.

Temperature / °C	*Rate of reaction / min^{-1}*
0	0.00
15	0.07
25	0.12
35	0.25
45	0.50
55	0.28
65	0.00

(a) Plot the data on a graph. **[4]**
(b) What is the optimum temperature of this reaction? **[1]**
(c) Describe the shape of the graph between 10 and 40 °C. **[2]**
(d) Calculate the rate of increase between 20 and 30 °C. **[2]**

4.19 Figure 4.1 shows the effect of substrate concentration on an enzyme-catalysed reaction. Explain why the rate is increasing between points A and B and then levels off at point C. **[3]**

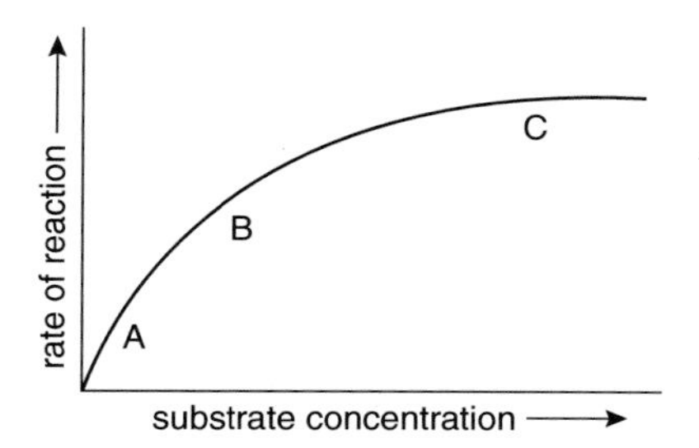

Figure 4.1

4.20 Describe how you would carry out an experiment to investigate the effect of pH on the activity of the enzyme catalase in yeast cells. **[4]**
What precautions would you have to take? **[3]**

4.21 Figure 4.2 shows the activity of the commercial enzyme Alcalase® at different pH values. Alcalase® is a protease enzyme.

(a) What compounds are digested by this enzyme? **[1]**

(b) Describe the change in enzyme activity with pH. **[3]**

(c) How does this curve compare to the pH curve of a human digestive enzyme such as pepsin? **[2]**

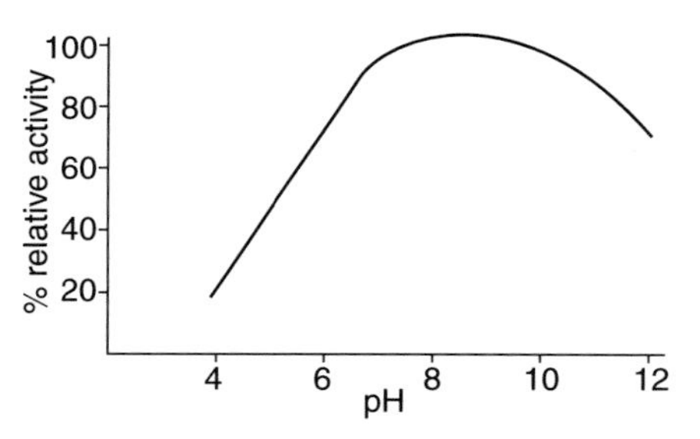

Figure 4.2 Data source: Novo Nordisk Bioindustries

4.22 The enzyme amyloglucosidase hydrolyses the $\alpha 1,4$ and $\alpha 1,6$ links in starch, releasing glucose. A solution of the enzyme was mixed with sodium alginate solution. The resulting mix was drawn up into a syringe and then dropped slowly into a calcium chloride solution to produce alginate beads. The beads were drained and placed in a column. Buffered starch solution was poured through the column. Glucose solution was collected at the bottom of the column.

(a) Why was the starch solution buffered? **[1]**

(b) What are the advantages of using an immobilised enzyme compared with an enzyme in a solution? **[2]**

(c) How could you test that the resulting solution contained glucose? **[2]**

(d) How could you measure the quantity of glucose that was produced? **[2]**

(e) What precautions would you take when carrying out this experiment? **[4]**

(f) How could you increase the amount of product? **[2]**

5 Heterotrophic nutrition

You should know that:

- digestion is the breakdown of large insoluble food molecules into small soluble molecules that can be absorbed
- food is mechanically digested by teeth and muscular movements of the wall of the stomach and intestine
- food is chemically digested by enzymes in the mouth, stomach and small intestine
- food is moved by peristalsis
- absorption of digested food takes place in the small intestine.

KEY FACTS

- Heterotrophs gain their food by eating plants or other heterotrophs. Most animals are holozoic since they feed on food from the bodies of other organisms, either plant or animal. Fungi and bacteria are saprotrophic or saprobiontic as they feed on organic material locked up in the bodies of dead plants and animals. Parasitic organisms feed on organic material taken from another living organism called a host. Mutualism is an association between two organisms where both partners benefit.
- The stages of digestion are ingestion, digestion, absorption, assimilation and egestion.
- Food is moved along the alimentary canal by peristalsis – the rhythmical contraction and relaxation of muscles.
- The wall of the alimentary canal typically consists of three main layers – outer muscle layer with circular and longitudinal muscles, submucosa and mucosa.
- Food is mechanically digested by teeth and muscles. Chemical digestion is carried out by hydrolytic enzymes which hydrolyse polymers into monomers.
- Starch is digested by salivary amylase in the mouth and pancreatic amylase in the small intestine. Disaccharides are digested by sucrase, lactase and maltase which are released from the epithelium of the ileum.
- Protein is digested by proteases. There are two groups of proteases: endopeptidases and exopeptidases. Endopeptidases (pepsin in the stomach, trypsin and chymotrypsin in the duodenum) hydrolyse peptide bonds within the protein molecule, creating shorter polypeptides. Exopeptidases hydrolyse the peptide bonds at the end of the polypeptide chains (aminopeptidases, carboxypeptidases and dipeptidases in the duodenum and ileum).
- Lipids are digested by lipase released from the pancreas, aided by bile salts in bile.

- Absorption takes place in the ileum. The villi increase the surface area. Uptake is further maximised by microvilli on the villi, a thin epithelial membrane and a good blood supply.
- Uptake of amino acids and monosaccharides occurs by diffusion, active transport and facilitated diffusion. Fatty acids and glycerol diffuse into the epithelial cells where they are recombined to form triglycerides which pass into lacteals.
- Release of the digestive secretions is controlled by nervous stimulation and hormones. The presence of food in the stomach causes the release of gastrin which stimulates gastric glands to release enzymes. The presence of acidic chyme in the duodenum causes the release of secretin and cholecystokinin which stimulate the liver and pancreas to release digestive secretions.
- Absorbed foods are carried by the hepatic portal vein to the liver where the food is assimilated and regulated.
- The colon absorbs water and mineral ions.
- Undigested food, together with dead cells, bacteria and mucus is egested as semi-solid faeces.
- The alimentary canal is adapted to diet. Herbivores have long guts for the digestion of cellulose, or a rumen containing mutualistic cellulose-digesting bacteria; carnivores eat a protein-rich diet, which is easy to digest so their gut is much shorter.

Questions for you to try

5.1 Identify the structures labelled A–H on Figure 5.1 which shows the human alimentary canal and some of the associated organs. **[8]**

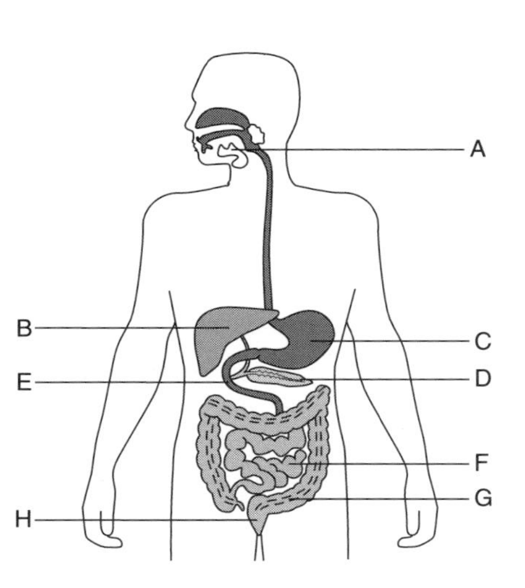

Figure 5.1

5.2 Which blood vessel carries food away from the intestine to the liver? **[1]**

5.3 How is food pushed along the small intestine? **[3]**

5.4 What is the difference between mechanical and chemical digestion? **[2]**

5.5 What is the role of the colon in a human? **[2]**

5.6 Name the compounds produced by the digestion of a triglyceride. **[2]**

5.7 Figure 5.2 shows the outline of a low power plan of a villus in the ileum.

(a) Complete the low power plan of the villus. **[4]**

(b) Label the following structures on your diagram: **[2]**

epithelium   capillary   submucosa   circular muscle

(c) How is the villus specialised for the uptake of digested foods from the ileum? **[4]**

(d) What is the function of the muscle layers in the ileum? **[1]**

Figure 5.2

5.8 Draw a simple flow chart to show all the stages and enzymes involved in the digestion of carbohydrates. **[6]**

5.9 The table below shows some of the enzymes which are involved in the digestion of proteins. Write the most appropriate word in the gaps. **[9]**

Enzyme	*Site of production*	*Site of action*	*Substrate*	*Product*
rennin				casein
	pancreas			polypeptides
carboxypeptidase		duodenum		

5.10 Pepsin is present in the stomach. Describe the conditions which are required in order for this enzyme to function. **[2]**

5.11 What is the role of gastrin in digestion? **[2]**

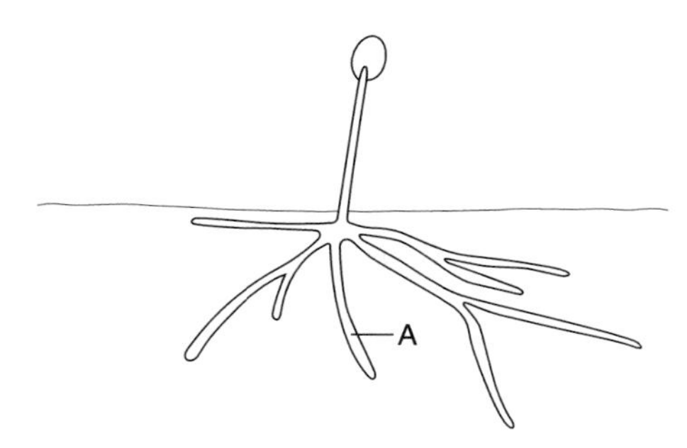

Figure 5.3

5.12 Figure 5.3 shows the pin mould *Rhizopus* growing across bread.

(a) Label the structure A. **[1]**

(b) How does this fungus obtain its food? **[2]**

(c) What is this mode of nutrition called? **[1]**

(d) How does this mode of nutrition differ from that of a parasite? **[2]**

5.13 Where is bile made? **[1]**

What is the main constituent of bile and what is the role of bile in digestion? **[4]**

5.14 Two groups of enzymes digest proteins. They are called endopeptidases and exopeptidases. Explain what these two groups of enzymes do. Which group is secreted first and why? **[6]**

5.15 The products of digestion have to be absorbed across the wall of the ileum and into the bloodstream. Describe the mechanism by which amino acids are transported. **[2]**

5.16 A number of hormones control digestive enzyme secretion. Figure 5.4 summarises the action of several hormones.

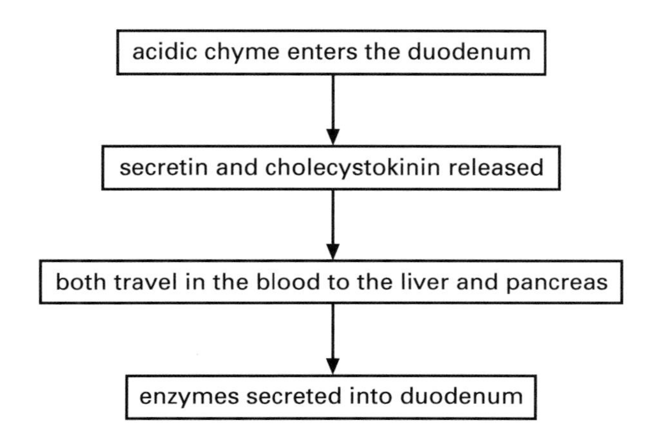

Figure 5.4

(a) Which part of the duodenum releases secretin and cholecystokinin? **[1]**

(b) Describe the effect of these two hormones on the pancreas. **[2]**

(c) What effect do these hormones have on the liver? **[2]**

(d) A third hormone, enterogastrone is released. What is its function? **[1]**

5.17 The bacterium, *Rhizobium*, is found in the root nodules of beans and peas. Describe the relationship between the bacterium and the plant. **[2]**

5.18 Ruminant animals, such as the sheep and cow, have a modified stomach called a rumen. Describe the processes that take place in the rumen. **[3]**

5.19 Produce a table that compares the features of the teeth of herbivorous and carnivorous mammals. How are they adapted to the mammal's diet? **[6]**

6 Exchanges with the environment

You should know that:

- gas exchange in unicellular organisms takes place over their entire surface
- humans have lungs with alveoli to increase the surface area available for gas exchange
- ventilation mechanisms involving the intercostal muscles and diaphragm draw air into the lungs
- the structure of the alveoli is designed to give efficient gas exchange
- inspired and expired air differ in their composition.

KEY FACTS

- Gas exchange surfaces have a large surface area and are permeable, thin and moist with a good transport system to maintain concentration gradients.
- Organisms with an internal gas exchange surface require a mechanism for ventilation to transport the gases from the environment to the exchange surface.
- Fish have gills which are well supplied with blood. Water is forced over the gills by the action of the mouth and operculum. Blood runs through the gills in the opposite direction to the flow of water, in a counter-current system to increase oxygen and carbon dioxide exchange.
- Insects have a network of tracheoles (tiny tubes) to carry oxygen directly to the cells. Spiracles lead to tracheae, which lead to tracheoles. A watery fluid can be removed from the tracheoles to increase the surface area for gas exchange. Larger insects can ventilate using their abdominal muscles.
- Mammals have a trachea leading to two bronchi which further divide into bronchioles. These end in alveoli, tiny sacs which greatly increase the surface area. They are surrounded by blood capillaries.
- In mammals, inhalation is brought about by intercostal muscles and the diaphragm.
- Ventilation rate is controlled by impulses from the respiratory centre in the brain and from receptors monitoring the changes in blood carbon dioxide levels.
- In plants, gas exchange takes place on the surface of the leaf mesophyll cells. Since their metabolic rate is lower than that of animals there is less demand for gases, so gases can diffuse into and out of the leaf, via the stomata.

Questions for you to try

6.1 Describe **four** common features of respiratory surfaces found in animals. **[4]**

6.2 How does gas exchange take place in a single-celled organism such as the *Amoeba*? **[2]**

6.3 Why is a specialised gas exchange surface required in a large multicellular animal? **[2]**

6.4 What does the term 'ventilation' mean? **[1]**

6.5 Figure 6.1 shows the respiratory system in a human. Label the structures A–G. **[7]**

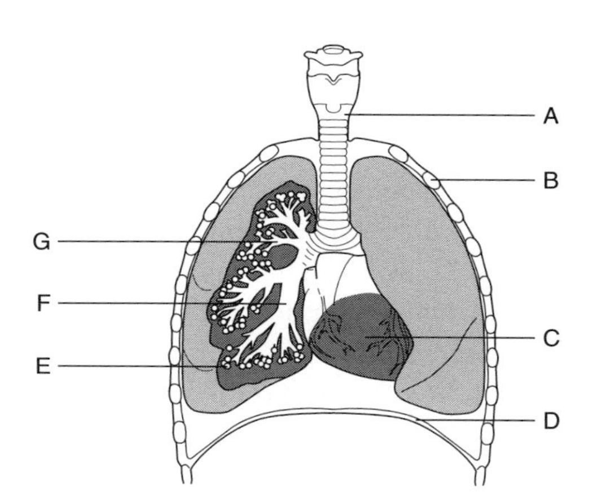

Figure 6.1

6.6 Complete the gaps in the following passage describing inspiration in a human:

At the start of inspiration, the ribs move upwards and ________ by the contraction of the ______ intercostal muscles. At the same time the diaphragm __________, causing it to become flatter. These movements _______ the volume of the thorax. The pressure within the thorax ________ and air rushes into the lungs. **[5]**

6.7 The table below shows the percentage composition of inspired, alveolar and expired air.

Gas	*Inspired air %*	*Alveolar air %*	*Expired air %*
nitrogen	79.0	80.7	79.6
oxygen	21.0	13.2	16.4
carbon dioxide	0.04	5.5	4.0
water	variable	saturated	saturated

Explain why:

(a) the percentage of nitrogen gas in expired air is virtually the same as in inspired air, **[1]**

(b) the percentage of oxygen is lower in alveolar air than in expired air, **[2]**

(c) expired air is saturated with water vapour. **[1]**

6.8 Describe the route taken by carbon dioxide to reach the cytoplasm of the palisade cells in a leaf. **[3]**

6.9 Describe how oxygen passes from air in the alveolus into the blood of a lung capillary in a human. **[3]**

6.10 Stomata can open and close. Describe the mechanism which causes changes in ion concentration within the guard cells and explain how this leads to a change in turgidity. **[4]**

6.11 Explain the terms 'tidal volume', 'vital capacity' and 'residual volume'. **[3]**

6.12 Figure 6.2 shows a spirometer tracing of an adult human.

(a) The tidal volume at rest on this spirometer reading is 500 cm^3. If the breathing rate is 11 breaths per minute what is the volume of air breathed in per minute? **[1]**

(b) Of the 500 cm^3, only about 350 cm^3 actually reaches the alveoli. What happens to the rest of this air? **[1]**

(c) What happens to the tidal volume once the person starts to exercise? **[1]**

(d) By how much does the tidal volume increase? **[1]**

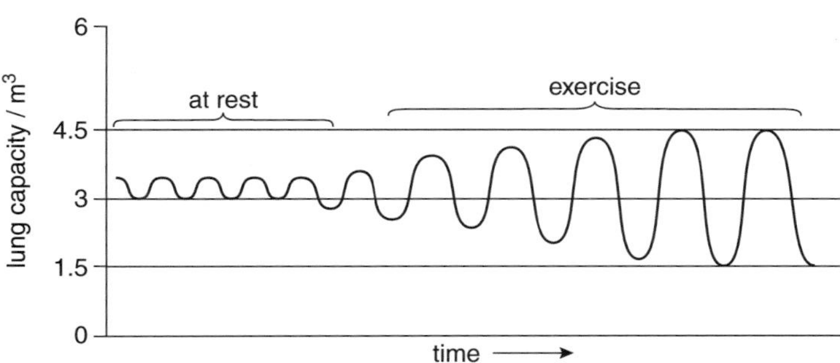

Figure 6.2

6.13 Maximum breathing capacity is the maximum amount of air that can be breathed in and out of the lung in a unit of time. This value can be increased by training. At the start of a training programme, a person's lung capacity was 120 litres min^{-1}. Ten weeks later this had increased to 150 litres min^{-1}. Calculate the percentage increase in lung capacity. **[2]**

6.14 **(a)** Describe the mechanism by which a fish moves water through its gills. **[3]**

(b) Mammals extract about 25% of the available oxygen from air in the lungs. Many fish extract up to 80% of the oxygen dissolved in water. Explain how the countercurrent flow in the gills increases the uptake of oxygen into the blood. **[2]**

6.15 Compare and contrast the method by which insects and humans obtain their oxygen. **[3]**

6.16 Breathing can be controlled voluntarily, but the medulla in the brain is responsible for automatic control.

(a) Describe how the inspiratory and expiratory centres in the medulla work together to control breathing. **[3]**

(b) How does the body monitor carbon dioxide concentration in the blood? **[2]**

(c) What effect does an increase in carbon dioxide concentration have on breathing? **[1]**

6.17 Large insects such as the locust can increase the ventilation of their tracheal system by contracting their abdominal muscles. Ventilation rate can be observed under different conditions using the simple apparatus shown in Figure 6.3.

(a) How could you use this apparatus to record the change in ventilation rate with changing carbon dioxide concentrations? **[2]**

(b) What precautions would you need to take? **[3]**

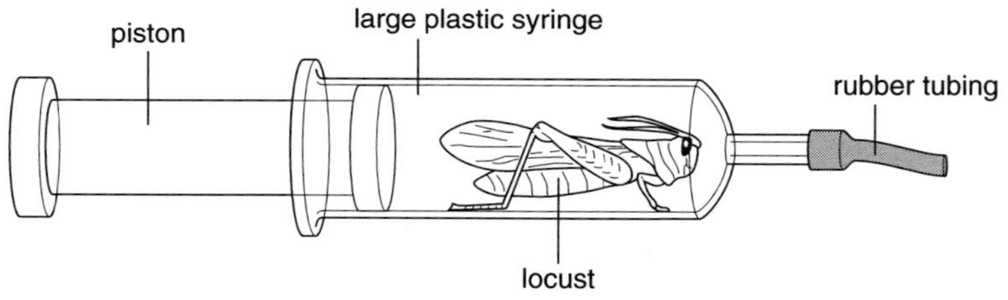

piston

large plastic syringe

rubber tubing

locust

Figure 6.3

6.18 **(a)** Describe **three** ways by which smoking damages the lungs. **[3]**

(b) Figure 6.4 shows the annual number of deaths from lung cancer, other cancers, bronchitis and tuberculosis of the lungs in men from 1916 to 1960. Compare the graphs for lung cancer and tuberculosis over this period. **[3]**

(c) What is the percentage increase in the annual number of deaths from lung cancer between 1940 and 1950? **[2]**

(d) Suggest reasons for the pattern of changes shown in the graphs for lung cancer and tuberculosis. **[2]**

(e) Explain the appearance of deaths from bronchitis in 1935. Why did the death rate stay at a relatively constant level? **[2]**

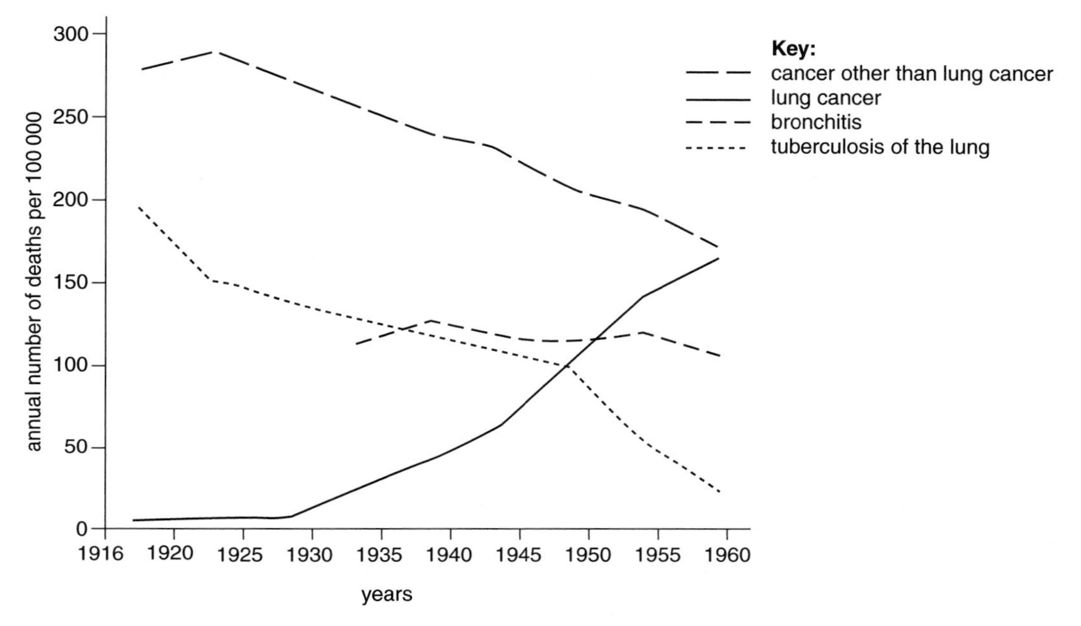

Figure 6.4

7 Transport

You should know that:

- large organisms need a transport system to move materials around their bodies
- blood consists of a liquid called plasma and three types of cell – red blood cells, white blood cells and platelets
- plasma carries materials such as amino acids, glucose and ions in solution
- red blood cells are full of haemoglobin which binds with oxygen; most of the carbon dioxide is carried in the plasma as hydrogen carbonate ions
- the heart is a muscular pump; it is divided into two parts, each with two chambers, which produces a double circulation so blood from the lungs is kept separate from the body circulation
- arteries carry blood away from the heart, veins carry blood to the heart and capillaries carry blood through the tissues – each of these blood vessels is adapted to suit their function
- water enters the roots of a plant by osmosis and evaporates from the leaves in transpiration
- water is carried through plants in the xylem vessels and sugars are moved through the phloem tissue.

KEY FACTS

- Humans have a double circulatory system consisting of the systemic circulation to the body and the pulmonary circulation to the lungs.
- The heart has four chambers (right atrium, right ventricle, left atrium, left ventricle).
- The heart is made up of cardiac muscle that can beat of its own accord. It is myogenic.
- The events of the cardiac cycle consist of systole (contraction of the atrium followed by contraction of the ventricle) and diastole (when the atrium and ventricle fill with blood).
- The sino-atrial node initiates electrical changes that bring about atrial systole and then ventricular systole.
- Arteries have thick, elastic walls and carry blood away from the heart under high pressure. Veins have thinner walls, a large lumen and semi-lunar valves. They carry blood back to the heart under lower pressure. Capillaries consist of a single layer of squamous epithelium so diffusion can occur.
- Blood consists of liquid plasma and blood cells.
- Plasma transports materials in solution and suspension.

- Red blood cells or erythrocytes are biconcave and lack a nucleus. They are packed with haemoglobin which binds to oxygen. The oxygen dissociation curve shows how the red blood cells become fully saturated with oxygen as it passes through the lungs, and then unload the oxygen in conditions of low partial pressure of oxygen in the tissues.
- Carbon dioxide is carried mainly in the plasma as hydrogen carbonate.
- In plants, xylem and phloem tissue are responsible for transport. Xylem carries water and mineral ions, phloem carries the soluble products of photosynthesis.
- Water moves through a plant as a result of water potential gradient. From roots to leaves, the water molecules move as a continual column as a result of cohesion and adhesion. Transpiration from the leaves is affected by wind, temperature and humidity.
- Minerals are carried by mass flow in the xylem. Minerals are taken into and out of cells by diffusion, facilitated diffusion and active transport.
- Sucrose, amino acids and other organic compounds are carried in the phloem. Sucrose is actively moved into the sieve tubes by transfer cells and then carried by mass flow around the plant.
- Xerophytic plants have smaller leaves or spines, hairs to trap water vapour, a thick waxy cuticle, and stomata in grooves or pits to reduce evaporation.

Questions for you to try

Transport in animals

7.1 Why do large organisms need a circulatory system? **[2]**

7.2 Mammals have a double circulation. What does this mean? **[2]**

7.3 Figure 7.1 shows the main blood vessels in a human body. Name the blood vessels labelled A–J on the diagram. **[9]**

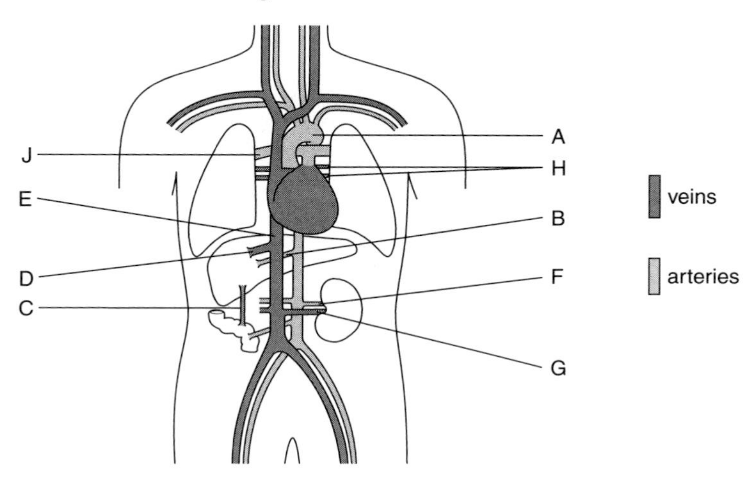

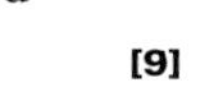

Figure 7.1

7.4 Figure 7.2 shows a cross section through the human heart.

(a) Label the structures A–G. **[7]**

(b) Why is the wall of the left ventricle much thicker than that of the right ventricle. **[2]**

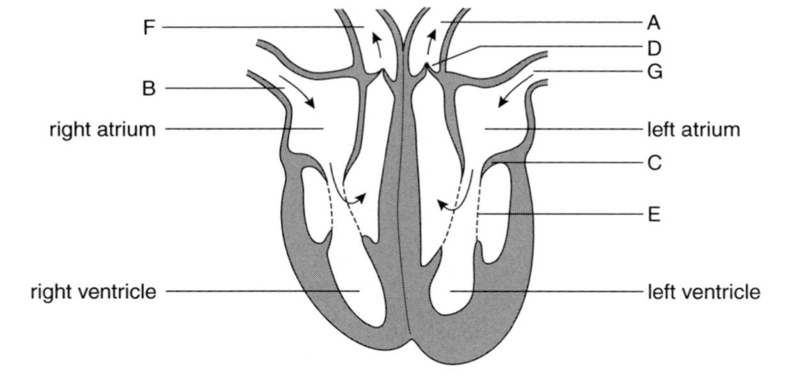

Figure 7.2

7.5 Cardiac muscle is described as being myogenic. What does this mean? **[2]**

7.6 Figure 7.3 shows a section through an artery and a vein.

(a) Measure the diameter of the artery as indicated on the photograph. If the magnification of the photograph is ×30, what is the actual diameter of the artery? **[2]**

(b) Using the photograph, compare the structure of the artery and vein. **[3]**

(c) How is the structure of the artery and vein related to their function. **[4]**

(d) What structure in the vein is not shown in this photograph? What is its role? **[2]**

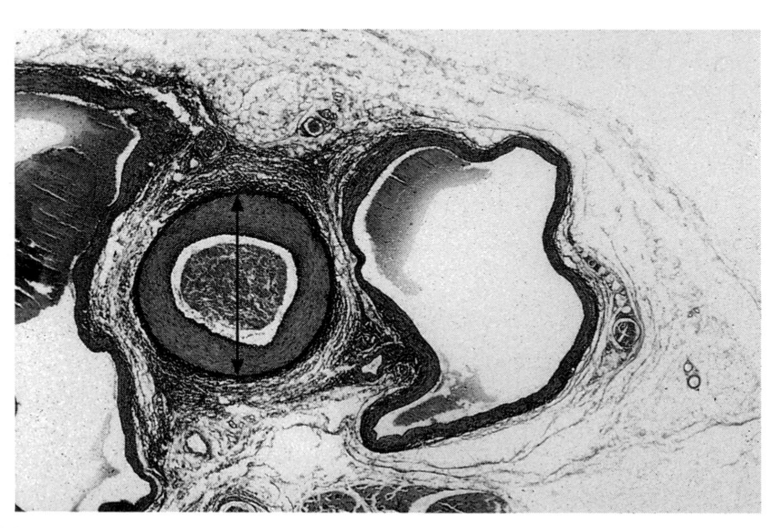

Figure 7.3

7.7 Figure 7.4 shows red blood cells (erythrocyte) and a lymphocyte. What is the role of:

(a) a red blood cell, **[1]**

(b) a neutrophil. **[1]**

(c) How is their structure linked to their function? **[4]**

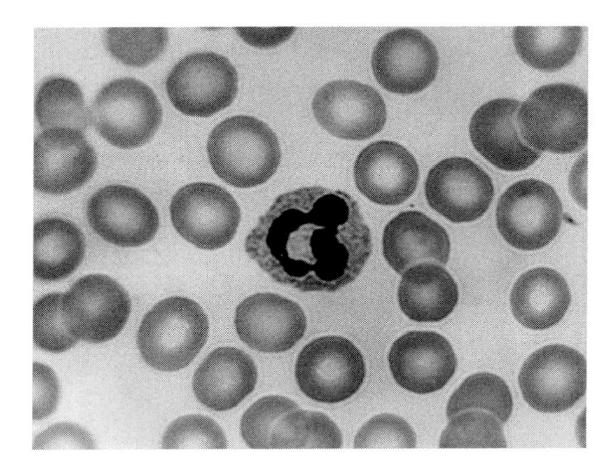

Figure 7.4

7.8 Distinguish between the terms plasma and tissue fluid. **[3]**

7.9 Using the information provided in Figure 7.5, describe the sequence of events that takes place during one cardiac cycle. Include in your answer the terms systole and diastole. **[6]**

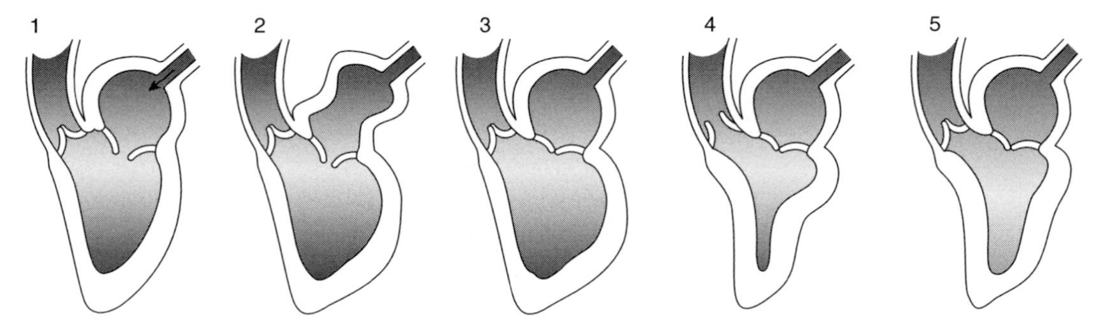

Figure 7.5

7.10 The sequence of events in the cardiac cycle needs to be carefully co-ordinated. Describe the role of the sino-atrial node (SAN) and atrioventricular node (AVN) in co-ordinating a heart beat. **[5]**

7.11 Figure 7.6 shows pressure changes to the left side of the heart and the aorta during the cardiac cycle.

(a) State what is happening at points A–D on the graph. Explain your answer. **[4]**

(b) If the time taken for one complete cardiac cycle is 0.8 seconds, how many cardiac cycles are there in one minute? **[2]**

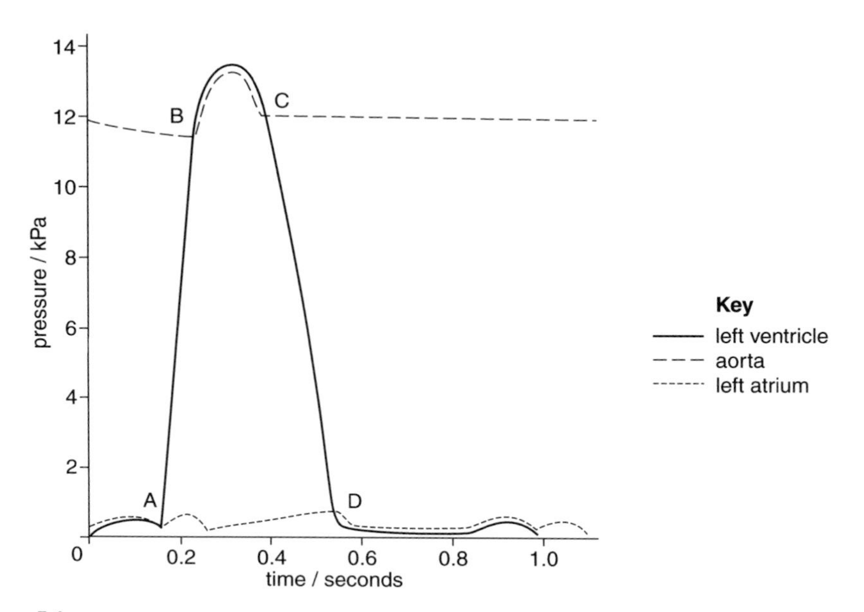

Figure 7.6

7.12 Figure 7.7 shows the oxygen dissociation curve for adult haemoglobin.

(a) Explain why the curve has an S shape. **[2]**

(b) Copy the graph and draw another curve to show:

(i) the oxygen dissociation curve under conditions of higher carbon dioxide concentration, **[1]**

(ii) the oxygen dissociation curve of fetal haemoglobin. **[1]**

(c) Explain your reasons for the shape and position of your curves. **[4]**

7.13 Figure 7.8 shows the increase in the red blood cell count of a group of mountaineers as they ascended a mountain.

(a) Describe the changes over the first four weeks. **[1]**

(b) Calculate the percentage increase in the red blood cell count over this period. **[1]**

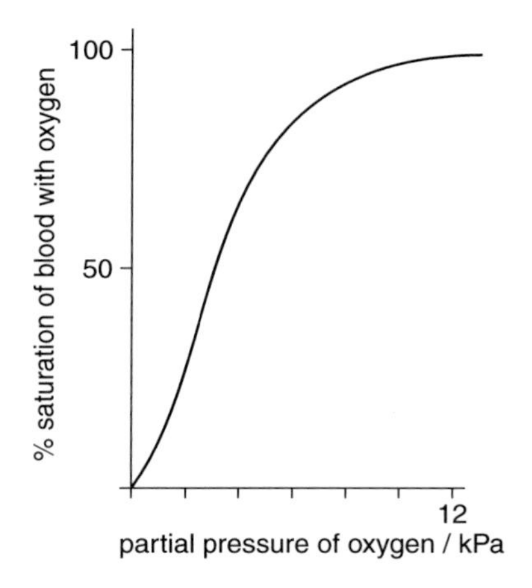

Figure 7.7

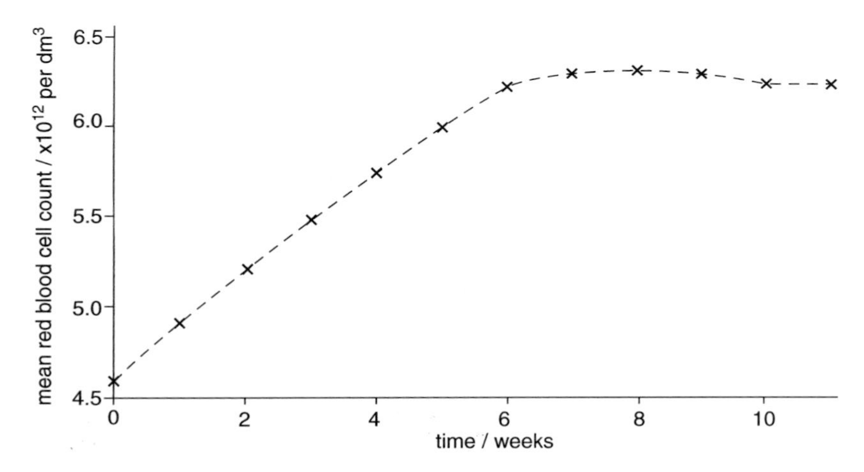

Figure 7.8

(c) It took the mountaineers two weeks to reach maximum altitude. What happened to the red blood cell count after they reached this altitude? **[1]**

(d) Explain why it is useful for the body to produce more red blood cells at high altitude. **[2]**

(e) In 1968, the Olympics were held in Mexico which lies at an altitude of just over 2000 m. Explain why many athletes trained at high altitude before going to these Olympics. **[2]**

Transport in plants

7.14 Draw a section across a plant root and another across a stem to show the distribution of xylem and phloem tissue, cambium and epidermis. **[4]**

7.15 By what process does water enter a root hair cell? **[1]**

7.16 By what process does water leave a leaf? **[1]**

7.17 How are xylem vessels adapted to carry water? **[3]**

7.18 Once within the root hair cell, water may move across a plant root following the apoplastic or symplastic pathways. Describe these two pathways. **[2]**

7.19 What is the role of the endodermis in a plant root? **[1]**

7.20 Explain how the forces of cohesion and adhesion are involved in the movement of water through the xylem vessel. **[4]**

7.21 Figure 7.9 shows the epidermis of a leaf with stomata.

(a) Calculate the stomatal density for this section of the leaf. **[3]**

(b) What is the role of stomata? **[2]**

(c) When are stomata normally open? **[1]**

(d) Stomata open when water enters the guard cells by osmosis. The mechanism which controls the opening and closing of stomata is unclear. Explain the possible role of potassium and hydrogen ions in the opening of stomata. **[4]**

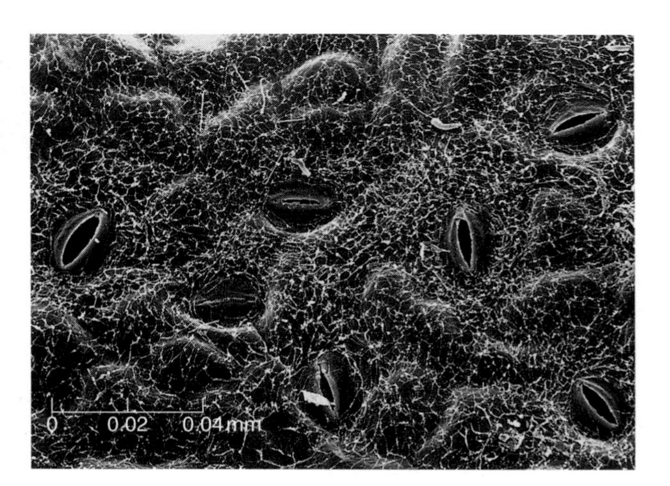

Figure 7.9

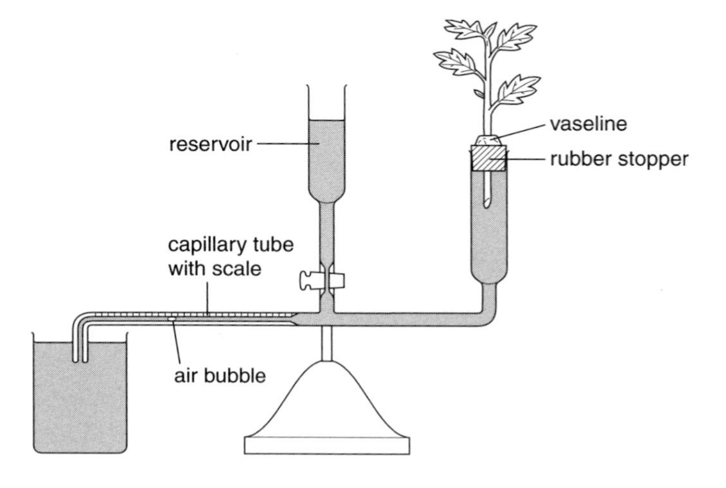

Figure 7.10

7.22 Potometers are used to measure the amount of water taken up by a cut shoot. Figure 7.10 shows how it is used.

(a) Describe how you would use this apparatus to investigate the effect of wind speed on water uptake. **[4]**

(b) What precautions would you take during this experiment? **[3]**

7.23 Figure 7.11 shows phloem tissue in a plant stem.

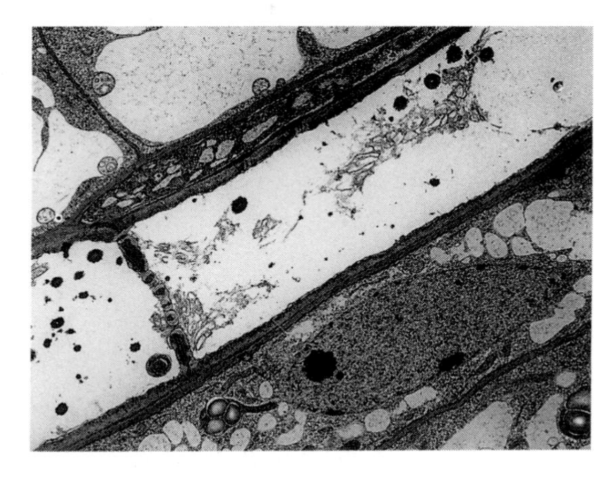

Figure 7.11

(a) Identify the sieve tube cell, sieve plate and the companion cell. **[3]**

(b) What is the role of the sieve tube cell? **[2]**

(c) What is the role of the companion cell? **[2]**

(d) There are many theories to explain how materials are moved along the sieve tube cells. The most likely is the mass flow or pressure flow hypothesis. Explain this theory. **[4]**

(e) Describe one way in which you could demonstrate the movement of materials in sieve tubes. **[3]**

7.24 Figure 7.12 shows a cross section of a xerophytic leaf

Describe **three** features which indicate that this plant lives in a dry climate. For each feature, explain how it helps the plant to survive in this environment. **[6]**

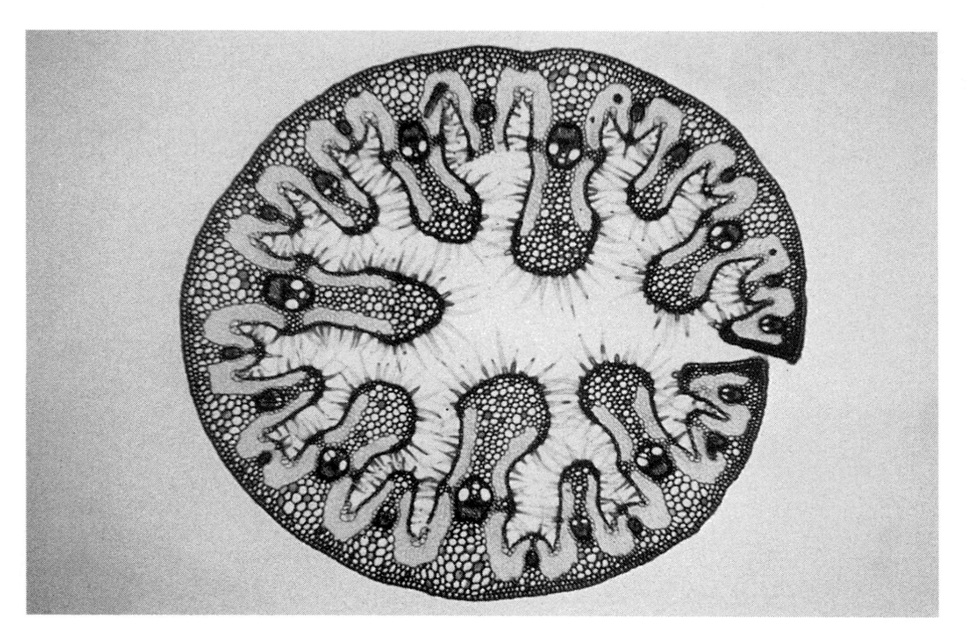

Figure 7.12

7.25 The results from an investigation into the rate of transpiration and absorption of water from a sunflower on a clear and hot summers day are shown in Figure 7.13

(a) Describe the changes in the rate of transpiration during the 24 hours of the experiment. **[3]**

(b) What caused these changes? **[2]**

(c) Compare the changes in the rate of transpiration to that of the rate of absorption. **[2]**

(d) If this pattern of transpiration and absorption continued for the next few days, what would happen to the plant? **[1]**

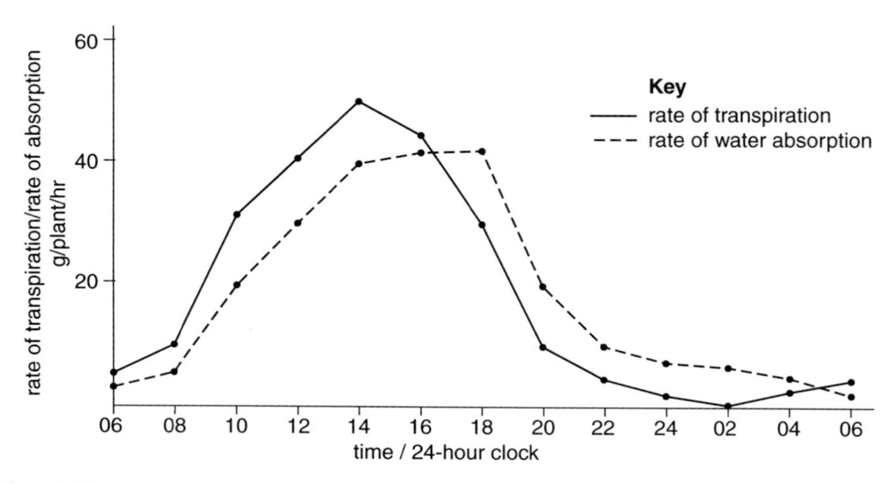

Figure 7.13

8 Nucleic acids, genes and gene technology

You should know that:

- DNA is made from two strands coiled to form a double helix, the strands are linked by a series of paired bases
- a gene is a section of the DNA molecule
- genetic engineering involves the transfer of one gene from a donor to a recipient using enzymes, to produce genetically altered organisms for the production of useful substances.

KEY FACTS

- DNA and RNA are polymers with a nucleotide monomer, each made up of a base, a pentose sugar and a phosphate group.
- DNA consists of two strands of nucleotides. The sugar is deoxyribose. There are four bases – adenine which pairs with thymine, and guanine which pairs with cytosine – linked together by hydrogen bonds.
- RNA is a smaller, single stranded molecule with ribose sugar and uracil instead of thymine. There are three types of RNA molecule – messenger, transfer and ribosomal RNA.
- DNA replicates by a semi-conservative method. The original strand of DNA splits and nucleotides from the cytoplasm attach themselves to the exposed bases.
- The order of bases along the DNA molecule codes for the order of amino acids in a polypeptide chain.
- A codon is a group of three bases. Each codon codes for a specific amino acid.
- During protein synthesis, the DNA code is transcribed onto a mRNA molecule that moves into the cytoplasm where it attaches to a ribosome. Here, the code is translated using tRNA molecules. Two codons at a time are translated. Each tRNA molecule carries a single anticodon and one specific amino acid. The anticodon of the tRNA pairs with the appropriate codon on the mRNA. A peptide bond forms between two adjacent amino acids.
- In genetic engineering, a specific gene is located and removed from the donor DNA using restriction endonucleases. It is inserted into the recipient DNA using DNA ligase. This creates recombinant DNA which may be inserted into a cell where the gene will be expressed and a product manufactured. For example, the gene of human insulin has been inserted into bacterial DNA and now cultures of bacteria make insulin in large amounts cheaply and of a pure standard. Bacterial plasmids (circular lengths of DNA) are the most common vector.

- Cloning produces large numbers of genetically identical organisms.
- The polymerase chain reaction (PCR) is a method used to make multiple copies of fragments of DNA. PCR amplifies a single fragment into millions of copies. PCR is used in forensic science to multiply a small amount of DNA in order to carry out further tests.

Questions for you to try

8.1 What do the letters DNA stand for? **[1]**

8.2 Where do you find DNA in an eukaryotic cell? **[1]**

8.3 Name the different types of RNA found in a cell. **[3]**

8.4 Figure 8.1 shows a section of DNA.

(a) Identify the parts labelled A, B and C. **[3]**

(b) There are four types of base in DNA. What are their names and how do they pair together? **[6]**

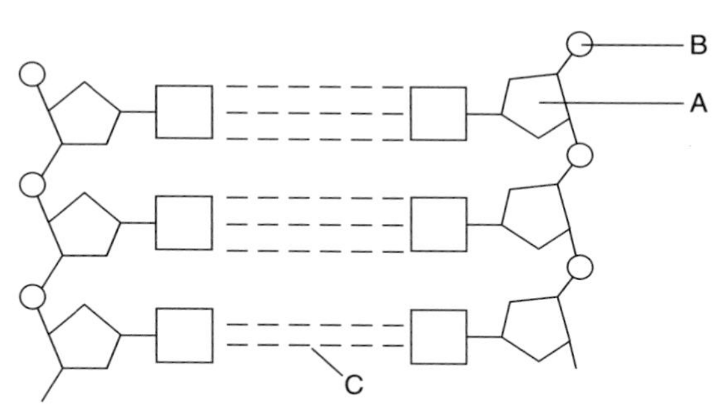

Figure 8.1

8.5 Produce a table which compares the structure of DNA and RNA. **[4]**

8.6 Explain how a strand of DNA undergoes semi-conservative replication. **[4]**

8.7 Meselson and Stahl carried out an experiment to prove that DNA replicated using the semi-conservative method. They grew bacteria on a medium of amino acids containing the nitrogen isotope, ^{15}N. The DNA was extracted from the bacteria and centrifuged. The DNA settled at a point in the tube which depended on the mass of the DNA. The bacteria were then moved to a medium containing ^{14}N and samples were taken from each generation of bacteria. The tubes containing DNA samples from three generations of bacteria are shown in Figure 8.2.

(a) How do these results prove that DNA undergoes semi-conservative replication? **[2]**

(b) What results would you expect if DNA replicated by a conservative method? **[2]**

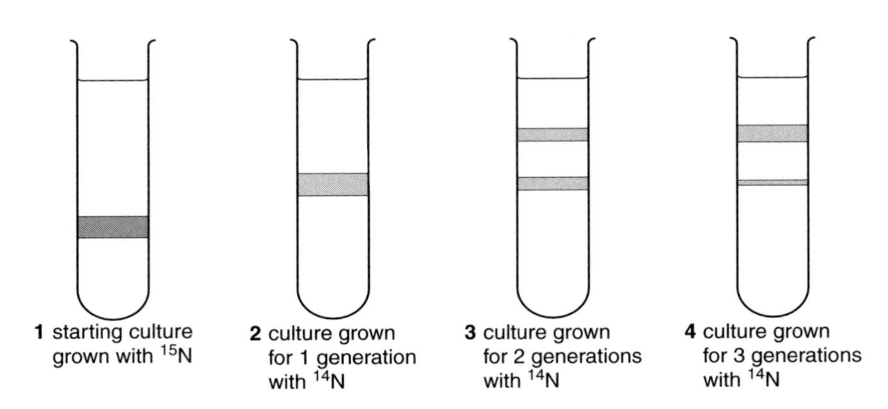

Figure 8.2

8.8 The genetic code is based on combinations of three bases, known as triplets or codons. What is the genetic code and why is the triplet important? **[5]**

8.9 What is the difference between a codon and an anticodon? **[2]**

8.10 Using the information provided in the table, write out the base sequences that would produce the following chain of amino acids for

(a) messenger RNA (mRNA), **[1]**

(b) DNA. **[2]**

glycine – isoleucine – valine – glutamic acid – glutamine

CUU* leucine	CCU proline	CAA glutamine	CGU argenine
AUU isoleucine	ACU threonine	AAU asparagine	GUU valine
GCU alanine	GAA glutamic acid	GGU glycine	UGG tryptophan

*messenger RNA codon corresponding to the amino acid

(c) Name the enzyme involved in the production of mRNA **[1]**

(d) The amino acids are assembled with the help of tRNA. Describe the structure of a tRNA molecule. **[2]**

8.11 What is an intron? **[1]**

8.12 Figure 8.3 shows the stages involved in the insertion of the gene for insulin into a bacterium.

(a) Name the substance that makes up the plasmid. **[1]**

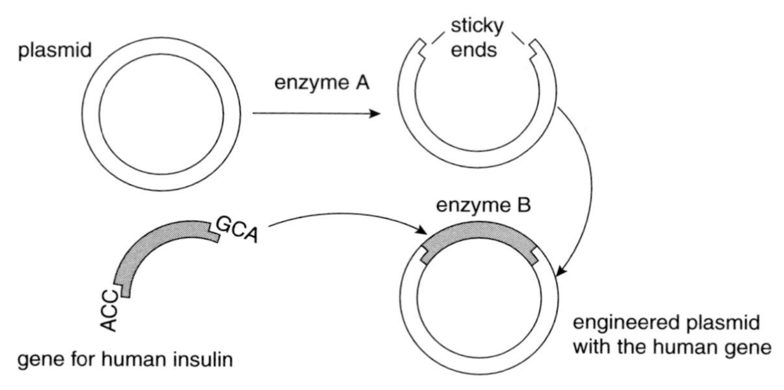

Figure 8.3

(b) Identify the enzyme labelled A. What is its role? **[2]**

(c) The sticky ends of the plasmid are complimentary to sticky ends on the gene for insulin. Write down the base sequences that are complimentary to those shown on the gene. **[1]**

(d) Identify enzyme B on the diagram. What is its role? **[2]**

(e) What term is given to a length of DNA formed from different sources? **[1]**

(f) How is the plasmid inserted into the bacterium? **[2]**

(g) How do scientists identify the bacteria which have taken up the plasmid? **[2]**

(h) What are the advantages of making insulin from engineered bacteria? **[3]**

8.13 The polymerase chain reaction (PCR) is used to make large quantities of identical DNA from very small samples. The flow chart summarises the steps involved in this process:

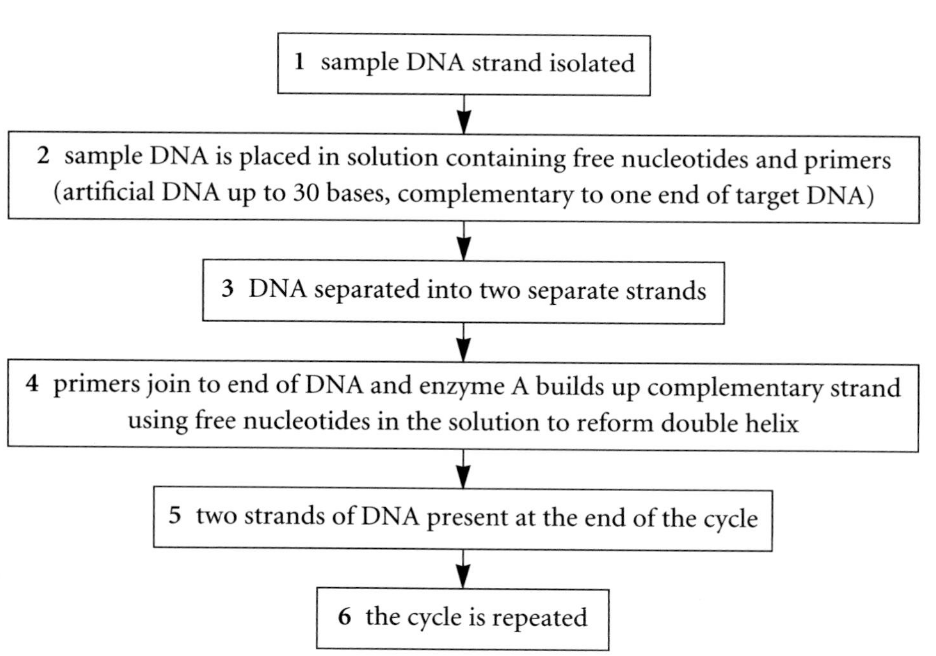

(a) There are two copies of the DNA present at the end of the first cycle. How many copies would you have after four cycles? **[1]**

(b) Identify enzyme A. **[1]**

(c) How does PCR differ from transcription? **[2]**

(d) Suggest how PCR may be important to forensic scientists. **[3]**

8.14 Write a short account of the processes that are involved in protein synthesis. **[10]**

9 Cell cycle

You should know that:

- there are 23 pairs of chromosomes in a human cell, with the exception of the gametes which have 23 single chromosomes
- in mitosis, the cell divides to produce two identical cells with the same number of chromosomes; mitosis produces new cells for growth and repair
- in meiosis, the division of the cell creates four cells, each with half the normal number of chromosomes; meiosis is used to produce haploid gametes during sexual reproduction, each gamete is different and this in turn produces variation.

KEY FACTS

- Human cells have 23 pairs of chromosomes in each body cell – this is the diploid number, made up of homologous pairs. The gametes have a haploid number of chromosomes, 23 chromosomes. Chromosomes are visible during nuclear division and may be displayed as a karyotype.
- The cell cycle comprises a period of non-division called interphase. This is made up of three separate phases, G1 – synthesis of cell components, S – the replication of DNA, and G2 – further growth and preparation, which leads up to the start of mitosis. Interphase may last hours or months, but mitosis itself takes less than one hour.
- Mitosis is used for growth and repair and asexual reproduction.
- In mitosis, each chromosome replicates, producing chromatids. The chromatids are separated during the division, so each daughter cell receives one copy of each chromosome. There are four phases – prophase, metaphase, anaphase and telophase.
- During meiosis, the homologous chromosomes are carefully separated, so that the gametes only have one of each pair of chromosomes. Meiosis halves the chromosome number, so contributes to genetic variation. Meiosis consists of two divisions. In the first division, the homologous chromosomes come together and are separated. This is followed immediately by a second division which is identical to mitosis. In the second division, the chromatids are separated, creating four cells each with the haploid number of chromosomes. Variation is created by crossing over between chromatids during division and from the independent assortment of chromosomes.
- Cancer is the result of uncontrollable cell division leading to the formation of tumours.

Questions for you to try

9.1 Name the main stages in mitosis. **[1]**

9.2 Figure 9.1 shows an animal cell during a mitotic division.

(a) Identify the structures A, B and C. **[3]**

(b) What is the role of structure C? **[1]**

(c) What stage of mitosis is shown in Figure 9.1? **[1]**

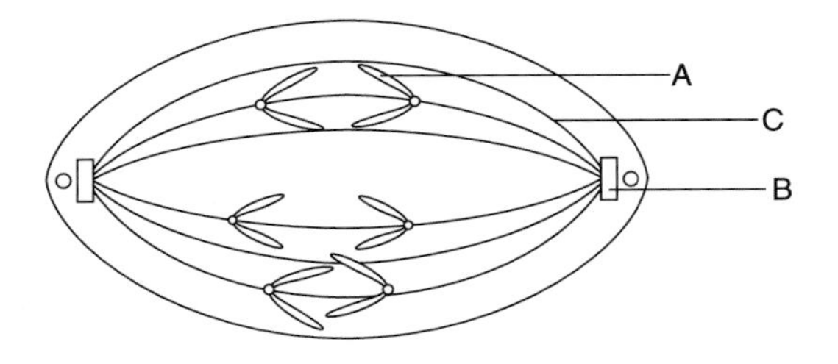

Figure 9.1

9.3 Distinguish between the terms haploid and diploid. **[2]**

9.4 How many chromosomes would you find in:

(a) a human cheek cell, **[1]**

(b) a human egg cell? **[1]**

9.5 The following table compares mitosis and meiosis. Copy the table and under each heading put either a tick if the statement is correct or a cross if it is incorrect. **[9]**

Feature	*Mitosis*	*Meiosis*
chromosomes replicate		
involves two nuclear divisions		
occurs in a haploid cell		
there is crossing over between the chromatids		
homologous chromosomes pair up		
sister chromatids are separated		
occurs during growth		
daughter cells are identical to the parent cell		
occurs during gamete formation		

9.6 What is meant by the term 'homologous chromosomes'? **[2]**

9.7 State **two** functions of mitosis. **[2]**

9.8 Many plants reproduce asexually. What are the advantages of asexual reproduction compared with sexual reproduction? **[2]**

9.9 State **three** ways by which meiosis creates genetic variation. **[3]**

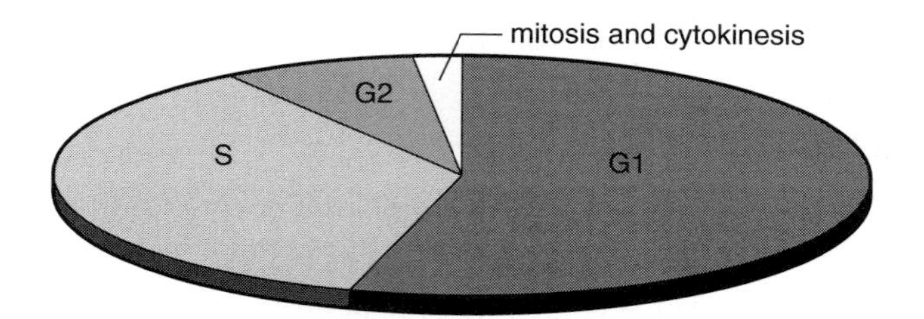

Figure 9.2

9.10 **(a)** Explain the meaning of the term cancer? **[2]**

(b) List **three** factors that are thought to cause cancer. **[3]**

9.11 Figure 9.2 represents the cell cycle

During which phase/s does the following take place:

(a) DNA replication, **[1]**

(b) energy production, **[1]**

(c) organelle replication? **[1]**

9.12 Write a brief account of the process of mitosis in an animal cell. **[8]**

10 Reproduction

You should know that:

- asexual reproduction produces large numbers of identical offspring
- sexual reproduction involves two separate sexes and produces variety amongst offspring
- flowering plants reproduce sexually by producing flowers containing stamens which produce the male gametes, and carpels which produce the egg cells
- pollination is the transfer of pollen from the anther to the stigma
- fertilisation is the fusion of the male and female gametes to produce a zygote which develops into a seed
- in animals, the males produce sperm, while females produce eggs
- in humans, fertilisation takes place in the oviduct; fetal development takes place in the uterus and birth occurs about 40 weeks after fertilisation.

KEY FACTS

- Asexual reproduction involves one individual and can produce large numbers of identical offspring by mitosis.
- Sexual reproduction involves the production of haploid gametes by meiosis followed by the fusing of male and female gametes to produce offspring which show genetic variation.
- Flowering plants produce flowers with sepals, petals, stamens (male parts) and carpels (female parts). Stamens consist of anthers where the pollen grains are produced. The male gamete is found within the pollen grain. The carpel consists of the stigma, style, ovary and ovule. The egg cell is formed in the ovule.
- Pollination is the transfer of pollen from the anther of one flower to the stigma of another, carried by the wind or animals.
- The pollen grain germinates to form a pollen tube which grows down into the ovule, allowing the male nuclei to enter the ovule.
- Double fertilisation occurs when one male nucleus fertilises the egg cell, while another fertilises the diploid endosperm nucleus.
- Spermatozoa are the male gametes of mammals. They are produced in the testes under the influence of follicle stimulating hormone (FSH) and testosterone. Spermatozoa are carried in a fluid called semen. Semen is ejaculated into the female vagina through the penis during copulation.
- The ovum is the female gamete and it is produced in the ovaries under the influence of FSH and luteinising hormone (LH).
- The human menstrual cycle lasts 28 days. During the first part of the cycle, FSH, LH and oestrogen are released. Ovulation occurs on day 14 when a

secondary oocyte (produced by meiosis) is released from the ovary. It travels down the oviduct where fertilisation may take place. The empty follicle produces progesterone. If fertilisation does not occur, the corpus luteum breaks up, progesterone levels fall and the lining of the uterus is shed, causing menstruation.

- If spermatozoa are present, the secondary oocyte completes meiosis, producing a haploid ovum. Fertilisation occurs to produce a zygote.
- The zygote divides repeatedly by mitosis forming a ball of eggs which implants in the wall of the uterus.
- Pregnancy in humans lasts about 40 weeks. At the end of pregnancy, oxytocin is released from the brain causing uterine muscle contractions which expel the baby. Prolactin causes the release of milk.

Questions for you to try

10.1 What does 'asexual reproduction' mean? Give two examples of asexual reproduction in animals. **[3]**

10.2 Compare and contrast the structure of an egg with that of a spermatozoan. **[3]**

10.3 Why do animals generally produce more spermatozoa than eggs? **[1]**

10.4 What is a clone? How is a clone produced? **[2]**

10.5 Figure 10.1 shows a generalised flower.

(a) Identify the structures labelled A–F. Give one function for each structure. **[6]**

(b) Give two reasons why this flower is insect pollinated. **[2]**

(c) Give two ways in which the structure of a wind pollinated flower, such as a grass, would differ from this flower. **[2]**

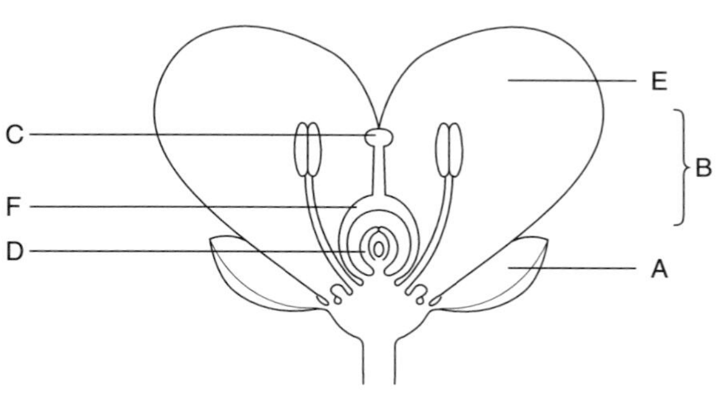

Figure 10.1

10.6 Describe the process of tissue culture in plants. Why is tissue culture useful? **[5]**

10.7 **(a)** Copy and complete the boxes in the generalised life cycle shown in Figure 10.2. **[3]**

(b) Indicate where mitosis and meiosis would take place. **[2]**

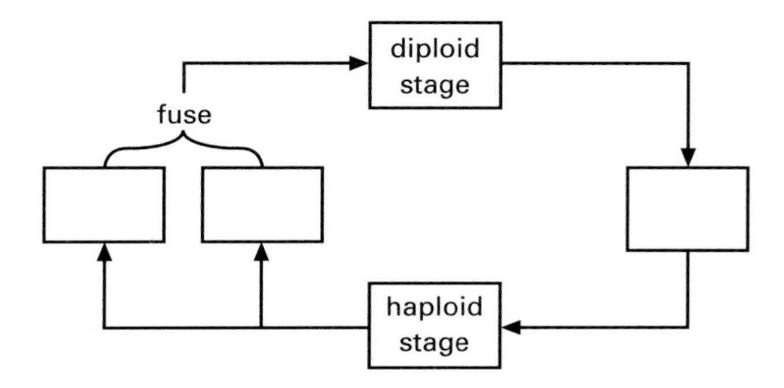

Figure 10.2

10.8 Distinguish between the terms protogyny and protandry. **[2]**

10.9 Once a pollen grain lands on a stigma it germinates and produces a pollen tube that grows through the style to the ovary. Describe the events that take place from this point to fertilisation. **[3]**

10.10 Double fertilisation is unique to flowering plants. Explain what this term means. **[2]**

10.11 What processes take place when a seed germinates? **[3]**

10.12 Figure 10.3 shows the male reproductive system. Identify the structures labelled A–G. **[7]**
Why does structure F hang outside the body? **[2]**

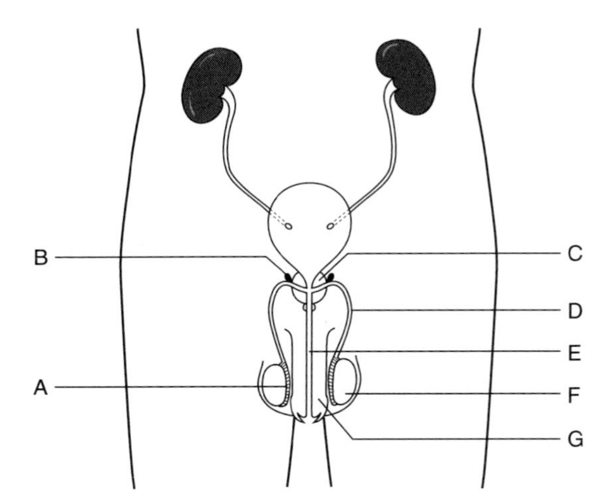

Figure 10.3

10.13 Copy the table below and indicate with ticks the correct statements for each hormone. **[8]**

	FSH	*LH*	*Oestrogen*	*Progesterone*
secreted from the pituitary				
repairs the endometrium				
inhibits the release of LH by negative feedback				
brings about formation of the corpus luteum				
secreted by the corpus luteum				
stimulates the development of a follicle				
brings about ovulation				
maintains the endometrium				

10.14 **(a)** What features of the placenta aid the transfer of materials between the mother and the fetus? **[2]**

(b) Why does maternal blood never come into direct contact with fetal blood? **[2]**

(c) Name **two** substances that enter the umbilical vein and two substances that leave from the umbilical artery. **[2]**

(d) What is the role of the amniotic fluid? **[1]**

10.15 Figure 10.4 shows, in outline, the process of spermatogenesis. Identify the cells labelled A–E. For each cell indicate whether it is diploid (2n) or haploid (n). **[5]**

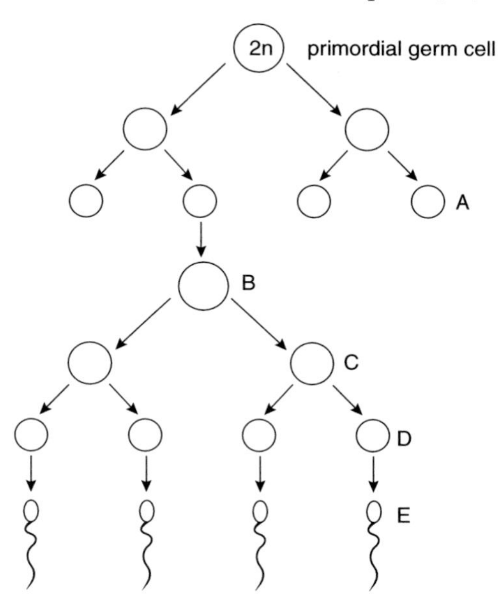

Figure 10.4

11 Genetics

You should know that:

- genes exist in different forms called alleles
- there are dominant and recessive alleles
- homozygous means the alleles are the same and heterozygous means that the alleles are different
- the genotype is the genetic make up of a person and phenotype is the appearance of an individual
- monohybrid crosses study the inheritance of a single gene – if two pure breeding or homozygous parents are crossed and their offspring allowed to interbreed the resulting F_2 generation will have the ratio of 3:1 dominant to recessive
- in humans, females have XX sex chromosomes and males carry XY sex chromosomes
- some genetic diseases are sex-linked, for example haemophilia.

KEY FACTS

- A gene is a length of DNA that codes for a particular characteristic.
- The genotype is the genetic make-up of an organism.
- The phenotype is the expression of the genetic makeup and it is affected by the environment.
- Alleles at a particular locus (particular position on the chromosome) may be homozygous (the same) or heterozygous (different).
- A dominant allele will express itself in a heterozygous state and will mask the recessive allele. Recessive alleles are only expressed in the homozygous state.
- Codominant alleles code for different proteins and neither dominates the other. The heterozygote has a phenotype which is a combination of both alleles, for example sickle cell anaemia, MN blood grouping.
- There may be multiple alleles of the same gene, for example A, B, O blood groups.
- Monohybrid crosses consider a single characteristic. The ratio for the phenotypes from a cross between two heterozygotes is 3:1 dominant to recessive.
- Dihybrid crosses study the inheritance for two characteristics. The ratio of phenotypes from a cross between two heterozygotes is 9:3:3:1.
- Gene interaction causes a phenotype to be produced by two or more pairs of alleles acting together, for example the shape of the comb in domestic chickens.
- Some genes occur on the same chromosomes and are inherited together. These genes are linked and the resulting ratio for two linked genes will be 3:1 rather than 9:3:3:1.

- Some genes are carried on the female X chromosome. These genes are sex linked because they are inherited with the X chromosome. These genes are not present on the Y chromosome so recessive sex-linked alleles are likely to be expressed in the male.

Questions for you to try

11.1 Write sentences to explain each of the following terms:
allele, dominant, recessive, homozygous, heterozygous, codominant. **[12]**

11.2 Homozygous purple-stemmed tomatoes were crossed with green-stemmed plants. The F_1 were all purple-stemmed. When the F_1 plants were allowed to self-pollinate the resulting F_2 produced 310 purple-stemmed plants and 120 green-stemmed plants.

(a) Which is the dominant allele? **[1]**

(b) Draw a genetic diagram to show the F_1 and F_2 crosses. **[5]**

(c) The F_1 plants were back crossed to a green-stemmed plant. The F_2 were 47 purple-stemmed and 55 green-stemmed plants.
Draw a second genetic diagram to show this back cross. **[4]**

11.3 Homozygous tall, white-flowered plants were crossed with homozygous short, red-flowered plants. Tall and white are dominant to short and red.

(a) What are the genotypes of the parent plants? **[2]**

(b) What is the genotype and phenotype of the F_1? **[2]**

(c) By means of a genetic diagram, show what happens when the F_1 plants are back-crossed to a recessive plant. Give the genotypes and the ratios of the phenotypes of the F_2. **[5]**

11.4 In dogs, dark coat colour (D) is dominant to albino (d), and short hair (H) is dominant to long hair (h). These two genes are not linked. A pure breeding, dark, short-haired dog is crossed with a pure-breeding albino long-haired dog.

(a) What is the genotype and phenotype of the F_1 puppies? **[2]**

(b) Two of the F_1 dogs are crossed and an F_2 produced. Draw a Punnett square to show the parental gametes and the genotypes and phenotypes of the offspring. What is the F_2 ratio of phenotypes? **[6]**

11.5 A plant with hairy stems and yellow flowers was crossed with a plant which had hairy stems and white flowers. Yellow flower colour is dominant to white. The seeds from the F_1 plants were sown and plants with the following characteristics were obtained:

28 plants with hairy stems and yellow flowers
35 plants with hairy stems and white flowers
10 plants with smooth stems and yellow flowers
11 plants with smooth stems and white flowers

Use the following key: H is hairy stems, h is smooth stems, Y is yellow, y is white

(a) Which is dominant – hairy stems or smooth stems? Why? **[2]**

(b) What is the genotype of the parents? **[2]**

(c) Draw a genetic cross to show the genotype and phenotype of the F_1 plants. **[3]**

(d) What is the ratio of hairy stems to smooth stems? **[1]**

(e) What is the ratio of yellow to white flowers? **[1]**

11.6 In an experiment, a homozygous tomato plant with a purple hairy stem was crossed with a homozygous tomato with a green, hairless stem. Both purple and hairy are dominant. The F_1 plants were allowed to self-pollinate to produce an F_2. The F_2 seeds were planted and the resulting phenotypes are shown below:

purple, hairy stem	150
purple, hairless stem	48
green, hairy stem	15
green, hairless stem	15

(a) What is the ratio of phenotypes in the F_2? **[2]**

(b) What was the expected ratio of phenotypes? Why? **[2]**

(c) Why do you think there is a difference between the observed and expected results? **[3]**

(d) How could you test that the differences were acceptable? **[2]**

11.7 In humans, I^A, I^B, and I^O are the three alleles of a gene that determines the ABO blood group. The gene is not sex linked.

AB individuals have the genotype I^AI^B
A individuals have the genotype $I^A\ I^A$ or $I^A\ I^O$
B individuals have the genotype $I^B\ I^B$ or $I^B\ I^O$
O individuals have the genotype $I^O\ I^O$
A woman with blood group O had three children. Their blood groups were A, B and O. She claims to have had only one partner. Is it possible for one man to have fathered all the children? Explain your answer. **[6]**

11.8 When red-flowered petunia plants are crossed with white-flowered plants, the resulting F_1 all have pink flowers.

(a) Explain how this is possible using genetic diagrams. **[3]**

(b) The F_1 plants are crossed to produce a F_2. Draw a genetic cross to show the genotypes and phenotypes of the F_2 plants. **[4]**

11.9 Red-green colour blindness is a sex-linked recessive condition. The gene for colour blindness is carried on the X chromosome. Figure 11.1 shows a family tree. Work out the genotype of the individuals labelled A–E. **[5]**

11.10 Some coat colours in cats are sex linked. Black coat colour is codominant to ginger. A cat that has one allele for black and one for ginger is tortoiseshell. The gene for this coat colour is carried on the X chromosome.

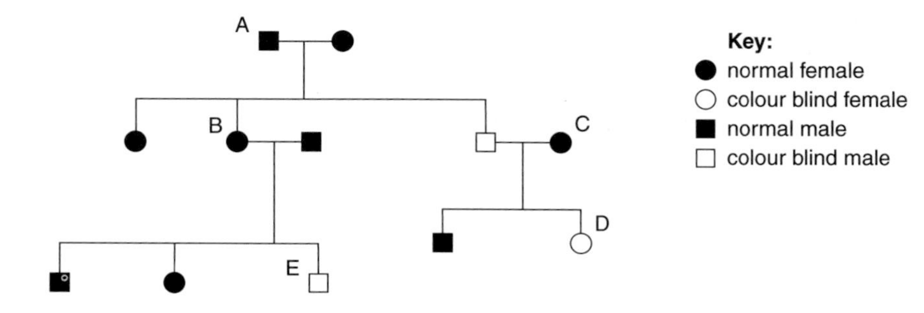

Figure 11.1

Describe the genotype and phenotype of the offspring of a cross between a pure breeding black female cat and a ginger male cat. **[4]**

11.11 In chickens, the shape of the comb is determined by two genes. There are four shapes of comb – pea, rose, walnut and single.

A pea-shaped comb is produced when there is at least one dominant allele for the pea shape (P) and recessive alleles for rose comb.
A rose comb is produced in the presence of at least one dominant allele for rose shape (R) and recessive alleles for pea.
A walnut comb is produced when there are dominant alleles for both pea and rose.
A single comb is a double recessive genotype.

Two pure breeding chickens are crossed, one with a pea comb (P) and the other with a rose comb (R).

(a) What are the parental genotypes? **[2]**

(b) What is the genotype and phenotype of the F_1? **[2]**

(c) The F_1 chickens are crossed to produce an F_2. Draw a Punnett square, showing the gametes of the parents, the genotypes and the phenotypic ratios of the offspring. **[6]**

11.12 **(a)** Explain what is meant by the term linkage. **[3]**

(b) If you were investigating the inheritance of two genes and you started with two pure breeding parents, what ratio of phenotypes would you expect in the F_2 if the genes were linked? Why? What would be the expected ratio if the genes were unlinked? **[4]**

12 Classification

You should know that:

- organisms have distinct features and can be grouped according to these features
- keys can be used to identify different organisms by looking at their visible features.

KEY FACTS

- Organisms that share many features are members of the same species. Individuals of the same species can breed together and produce offspring that can interbreed.
- Taxonomy is the categorising of living things in an orderly way and represents the ancestral relationships of these organisms.
- The binomial system of nomenclature gives each species two Latin names. The first name indicates the genus to which the organism belongs and the second name indicates the species. For example, the Latin name for humans is *Homo sapiens.*
- Classification follows a basic hierarchical system, originally devised by Linnaeus in the 18th century. Organisms are placed together in their taxa or groups. The largest group is the kingdom, which is divided into several phyla (singular phylum), classes, orders, families, genera and species.
- There are five kingdoms – Prokaryotae, Protoctista, Fungi, Plantae, Animalia.
- Prokaryotae comprise the single-celled prokaryotic organisms, such as bacteria.
- Protoctista comprise the single-celled eukaryotic organisms, such as *Amoeba.*
- Fungi are eukaryotic and may be single-celled or multicellular. Most comprise a network of hyphae with walls made of chitin and they reproduce by means of spores.
- Members of the Plantae are multicellular eukaryotic organisms that can photosynthesise. They have cells with cellulose cell walls, sap vacuole, and chlorophyll and other photosynthetic pigments.
- Members of the Animalia are eukaryotic organisms that feed heterotrophically.

Questions for you to try

12.1 What is the meaning of the term 'species'? **[2]**

12.2 Explain the term 'binomial system of nomenclature'. **[2]**

12.3 In classification, why is Latin used and not English? **[2]**

12.4 Copy the table below and insert the names for the five kingdoms used for classification purposes and for each kingdom give one feature and one representative example of the group. **[15]**

Kingdom	*Feature*	*Example*

12.5 The Latin name for human beings is *Homo sapiens.* Using this information and your knowledge of classification, fill in the gaps in the table below. **[4]**

Kingdom	
Phylum	
Class	Mammalia
Order	Primates
Family	Hominidae
	Homo
	Homo sapiens

12.6 The system of classification shown in 12.5 is described as hierarchical. What does this mean? **[2]**

12.7 What is the meaning of the term 'phylogeny'? How does a classification based on phylogeny differ from an artificial classification based on appearance? **[4]**

12.8 For the last 300 years, plants have been classified on the basis of their appearance. The flowering plants (Angiospermatophyta) were divided into two groups, the monocotyledons with one seed leaf and the dicotyledons with two seed leaves. Modern analysis of DNA sequences have show that flowering plants should be classified according to the type of pollen rather than the seed leaf. Many species that are new to science are being discovered in tropical rainforests. Using your knowledge of biology, suggest methods biologists can use to determine the correct classification of an unknown plant. **[3]**

13 Selection and evolution

You should know that:

- variation exists among individuals
- variation is caused by genes and the environment
- genes occur in alternative forms called alleles and these cause variation in inherited characteristics
- mutations can be caused by mutagens such as ionising radiation (gamma rays, ultraviolet light, X-rays) and environmental factors, such as tobacco smoke, cause changes in genes and chromosomes; some are harmful but others can be beneficial
- changes are spread by natural selection.

KEY FACTS

- Variation exists among members of the same species.
- Variation is a result of interaction between genetic and environmental factors.
- Meiosis can create variation through crossing over, independent assortment of chromosomes and random fertilisation.
- Variation can be continuous, as illustrated by polygenes, or discontinuous, as shown by single gene inheritance.
- The sum of all the genes in a population is called the gene pool.
- The frequency of the alleles in a population can be changed due to selection.
- The Hardy–Weinberg equilibrium can be used to estimate the frequency of alleles in a population.
- Selection can be directional, stabilising or disruptive.
- Mutations can affect genes and chromosomes.
- Gene mutations affect the sequence of bases on the DNA. For example, insertion (addition of one or more bases), deletion (loss of one or more bases) and substitution (switching of one base for another). Sickle-cell anaemia is an example of a substitution mutation.
- Chromosome mutations can be caused by non-disjunction when homologous chromosomes fail to separate properly during anaphase. This can cause polysomy (one extra chromosome) and polyploidy (one or more extra sets of chromosomes).
- Variation can lead to differential survival and reproductive success. Individuals with selective advantage are more likely to survive and reproduce and pass their genes on to the next generation. This process is called natural selection.

- Isolation can lead to speciation. Isolation mechanisms can involve geographical barriers such as rivers and mountains, behavioural or reproductive behaviour.
- Speciation can be allopatric or sympatric.
- Artificial selection is important in the development of domestic breeds of animals and new varieties of crop plants.

Questions for you to try

13.1 What is variation? **[2]**

13.2 Explain **two** ways in which meiosis contributes to genetic variation. **[2]**

13.3 State **five** ways in which humans can vary from each other. For each, indicate whether you think the variation is due to a gene, or environmental influence, or both. **[5]**

13.4 Define continuous variation and discontinuous variation. Draw two graphs to show the differences between the two types of variation and give an example of each. **[6]**

13.5 What is speciation? **[2]**

13.6 How does geographic isolation occur? **[3]**

13.7 **(a)** What is sympatric speciation? Give **two** examples to illustrate your answer. **[4]**

(b) How does sympatric speciation differ from allopatric speciation? **[2]**

13.8 The diagram below shows the codons along a length of RNA which code for six amino acids in a polypeptide chain.

CCU – ACG – CAG – GAC – GGC – AUG

(a) What would be the effect if there was a base substitution mutation in the second codon, causing the ACG codon to be changed to AAG? **[2]**

(b) What would happen if the C of this codon was deleted? **[2]**

13.9 Downs syndrome can be caused by non-disjunction. Explain the term non-disjunction and describe how it causes Downs syndrome. **[4]**

13.10 A mule is a cross between a horse and a donkey. The diploid number of chromosomes is 64 in a horse and 62 in a donkey.

(a) How many chromosomes would there be in a somatic (body) cell of the mule? Explain your answer. **[2]**

(b) Why is the mule infertile? **[2]**

13.11 Explain the following terms:

(a) gene pool, **[2]**

(b) gene frequency, **[2]**

(c) gene flow. **[2]**

13.12 Reproductive isolation can lead to speciation. What is the difference between prezygotic and postzygotic isolation? **[4]**

13.13 Studies of identical twins help researchers determine if traits are largely determined by genes or whether they are affected by the environment. The table below shows mean differences between twins, based on 50 pairs each of identical twins reared together, non-identical twins, non-twin siblings and 19 pairs of identical twins reared apart.

Trait	*Identical twins raised together*	*Identical twins raised apart*	*Non-identical twins*	*Non-twin siblings*
difference in height (cm)	1.7	1.8	4.4	4.5
difference in mass (kg)	1.9	4.5	4.6	4.7
difference in IQ score	5.9	8.2	9.9	9.8

Which of these three traits seems to be determined by a) genes and b) the environment? Explain your answers. **[6]**

13.14 Figure 13.1 shows the distribution of birth weights of 13 000 children born between 1935 and 1946 with the percentage mortality.

(a) Describe the distribution of birth weight. **[3]**

(b) What is the relationship between birth weight and mortality? **[2]**

(c) Suggest one disadvantage of having **(i)** a small and **(ii)** a large baby. **[2]**

(d) What type of selection is shown in this graph? **[1]**

13.15 The mottled beauty, *Cleora repandata*, is a moth that occurs in two forms, a light form and a dark form. Approximately, 10% of the moths collected from a pine forest in Scotland were found to be of the dark form. The moths spend the day on pine trunks, but at dusk move to another resting place. It was observed that the light moths were at an advantage during the day as they were well camouflaged against the bark of the pine trees, but when they flew at dusk they were more visible than the dark forms. The dark forms were more visible during the day, but almost invisible at dusk.

(a) Why is the percentage of dark forms much lower than that of the light forms? **[2]**

(b) What type of environmental change would suit the dark form? **[2]**

(c) The mottled beauty moth is an example of a polymorphism. What does this term mean? **[2]**
Give one other example of a polymorphism. **[1]**

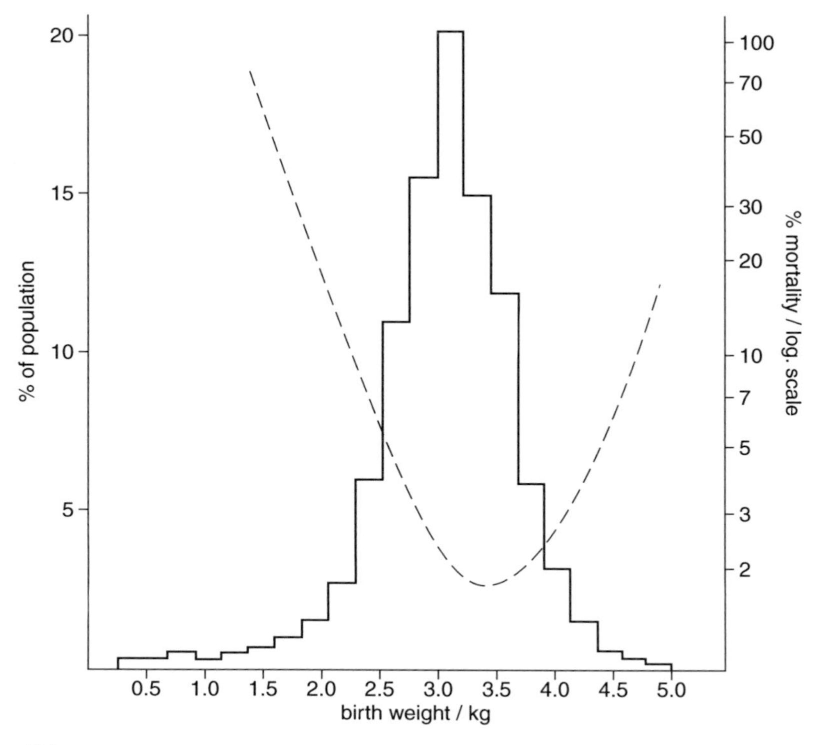

Figure 13.1

13.16 Biologists studied the variation in the amount of spontaneous movement which occurs when the fruit fly, *Drosophila*, is placed in standardised conditions and its activity recorded. The results are shown in Figure 13.2a.

(a) What type of distribution is present in this population of fruit flies? **[1]**

(b) The biologists took the flies with a low activity score and allowed them to breed amongst themselves. They did the same with those flies showing high activity. The results after 14 generations are shown in Figure 13.2b. Describe the differences between graphs a and b. **[2]**

(c) What type of selection is shown by these results? **[1]**

(d) Suggest the changes that might occur if the experiment was continued for more generations. **[2]**

(e) Suggest what might happen if the high activity flies produced by the breeding experiment were introduced to flies with a low activity score. **[2]**

13.17 The snail, *Cepaea hortensis*, is found in a variety of habitats such as woods, hedgerows and grasslands. The colour of the shell is variable. In woods, the shells tend to be heavily banded and predominantly brown. On grasslands, the shells tend to be predominantly yellow with light brown bands. Explain how selection can produce the differences between populations of the snail growing in different habitats. **[4]**

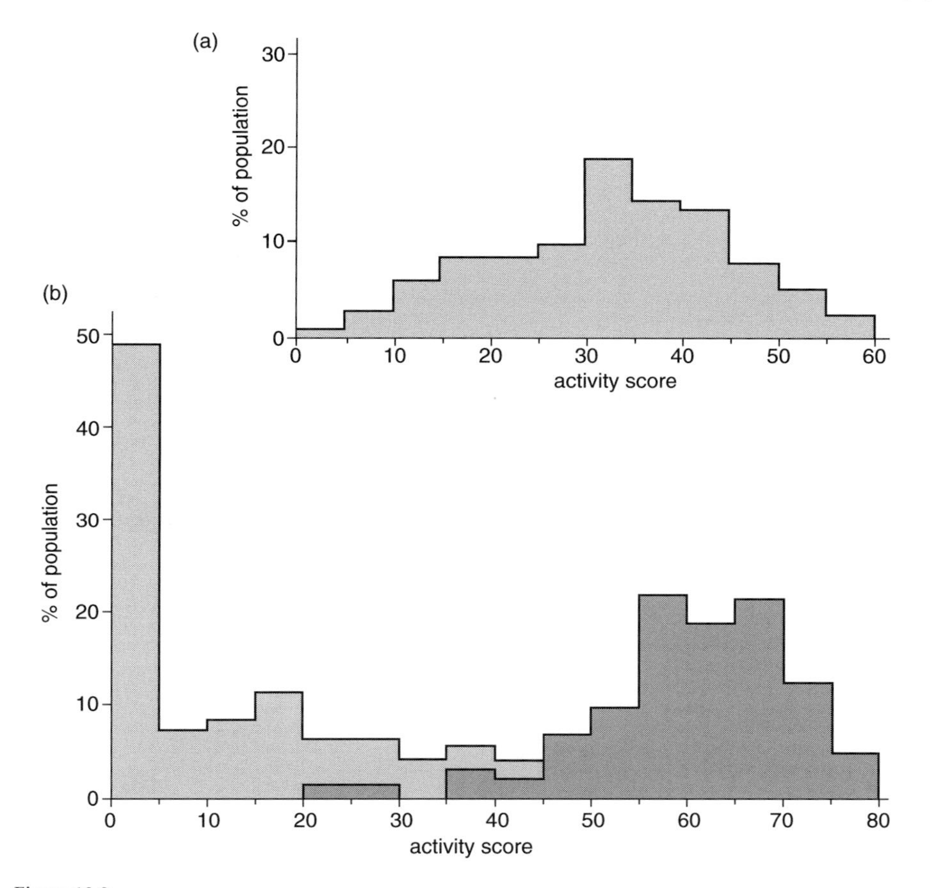

Figure 13.2

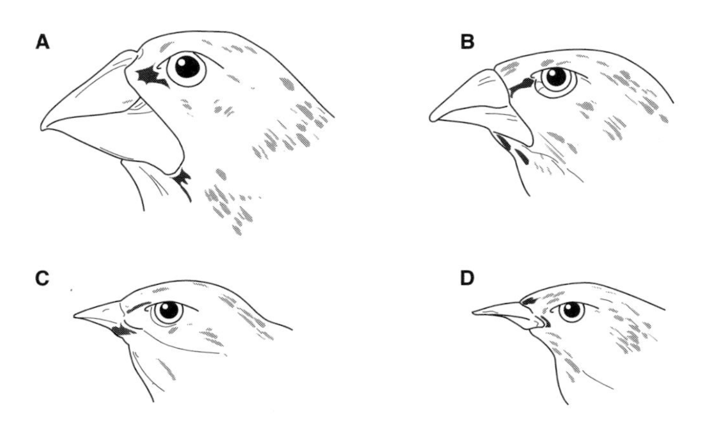

Figure 13.3

13.18 The Galapagos Islands lie approximately 1000 km off the coast of South America. There is one species of finch living on the mainland, but there are 12 species on the islands. The heads of four of the species are shown in Figure 13.3.

(a) Describe the differences in the shapes of the beaks. **[2]**

(b) How does a difference in beak shape affect the finch's way of life? **[2]**

(c) Why are there more species of finches on the island than on the mainland? **[2]**

13.19 Figure 13.4 shows how the ancestral wild pigeon has evolved into modern domestic varieties. Explain how pigeon breeders would have carried out this artificial selection. **[2]**

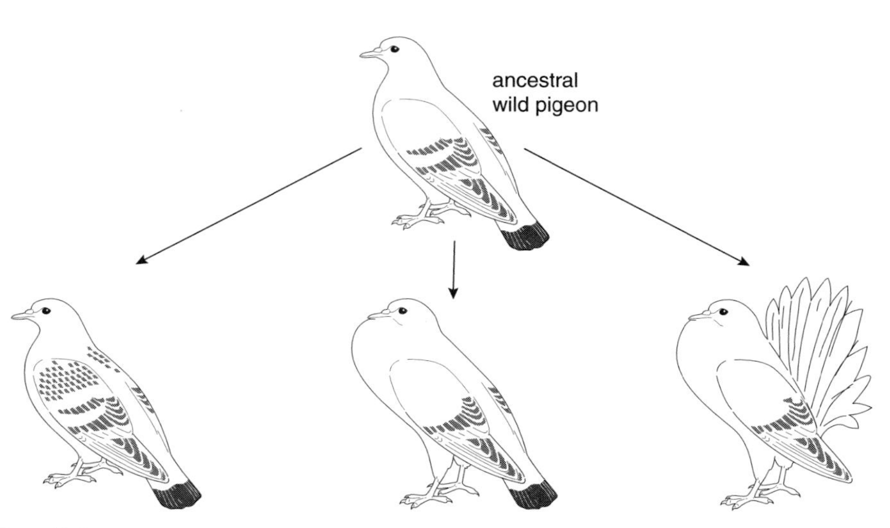

Figure 13.4

13.20 The peppered moth occurs in two forms, a dark melanic form and the normal non-melanic form. In a mark–release–recapture experiment in an industrial area the following results were obtained:

	Number captured, marked, released	*Number recaptured one week later*	*% recaptured*
melanic form	180	101	
non-melanic form	75	15	

(a) Copy and complete the table by calculating the figures in the percentage recaptured column. **[2]**

(b) Suggest **two** reasons for the difference in the percentage recaptured. **[2]**

13.21 Cystic fibrosis is a recessive genetic condition with a frequency of 1 in 2000 people. Calculate the frequency of the heterozygote. **[4]**

14 Respiration

You should know that:

- energy is released from glucose in cells by respiration
- aerobic respiration takes place in the presence of oxygen

$$\text{glucose} + \text{oxygen} \rightarrow \text{carbon dioxide} + \text{water} + \text{energy}$$

- anaerobic respiration takes place without oxygen, and releases less energy

$$\text{glucose} \rightarrow \text{lactic acid} + \text{energy}$$

KEY FACTS

- Adenosine triphosphate (ATP) makes energy available for chemical reactions within cells. ATP is hydrolysed to ADP, releasing energy. ADP can be phosphorylated to ATP by ATPase.
- Metabolic pathways such as respiration, are a series of enzyme-controlled reactions, involving enzymes such as oxido-reductases and hydrolases.
- Metabolism in a cell is the sum of all the catabolic and anabolic reactions. Catabolic reactions for example cellular respiration, involve the breakdown of substrates. Anabolic reactions, for example photosynthesis, involve the synthesis of new materials.
- Aerobic respiration requires oxygen, whereas anaerobic respiration takes place without oxygen.
- Aerobic respiration consists of glycolysis in the cytoplasm and Krebs cycle and electron transfer chain in the mitochondrion.
- In glycolysis, monosaccharides, such as glucose, are phosphorylated to form two 3-carbon molecules which are then converted to pyruvate with the formation of ATP and NADH.
- If oxygen is available, the pyruvate passes into the mitochondrion where it is converted to acetyl co-enzyme A, a 2-carbon compound, which enters the Krebs cycle. The acetyl chain combines with a 4-carbon compound, forming a 6-carbon compound which undergoes a series of dehydrogenation and decarboxylation reactions to reform the original compound, carbon dioxide, ATP and reduced co-enzymes (NADH and FADH).
- The hydrogen atoms enter the electron transfer chain, where the electron is removed and passed along a series of carriers. The energy released is used to form ATP. Each molecule of NADH yields three ATP. The final acceptor is oxygen, which combines with hydrogen to form water.
- The mitochondrion consists of a double membrane. The inner membrane is highly folded and is the site of the electron transfer chain. The matrix is the

site of the Krebs cycle. Highly metabolic cells have more mitochondria, for example, liver cells.

- If no oxygen is available, the pyruvate is converted to lactic acid (lactate) in animals or ethanol and carbon dioxide in plants and yeast. Anaerobic respiration yields much less ATP than aerobic respiration.
- The respiratory quotient (RQ) is the carbon dioxide produced divided by the oxygen used in a given time period. RQ values indicate the type of substrate being respired, for example, carbohydrates have a RQ of 1.

Questions for you to try

14.1 Name the following types of reaction and indicate what type of enzyme would be involved:

(a) a reaction in which carbon dioxide is removed, **[2]**

(b) a reaction in which hydrogen ions are removed. **[2]**

14.2 What is a hydrogen carrier? Give one named example. **[2]**

14.3 What is the product of anaerobic respiration in animals? **[1]**

14.4 What is the difference between a facultative anaerobe and an obligate anaerobe? **[2]**

14.5 Compare the yield of ATP in aerobic and anaerobic respiration. **[2]**

14.6 Reduction and oxidation take place in the electron carrier chain.

(a) Explain these **two** terms. **[2]**

(b) What is a redox reaction? **[1]**

14.7 Figure 14.1 shows a simple respirometer which is being used to measure the rate at which oxygen is taken up by germinating seeds.

(a) What is the role of the soda lime? **[1]**

(b) How would you estimate the volume of oxygen uptake by the seeds? **[2]**

(c) What precautions would you take to ensure you obtain accurate measurements? **[3]**

(d) How would you modify this apparatus to determine the amount of carbon dioxide produced by the seeds? **[3]**

(e) The RQ is calculated using the following formula:

$$\text{RQ} = \text{volume of carbon dioxide produced} \div \text{volume of oxygen used}$$

What does an RQ value of 1.0 tell you about the respiratory substrate? **[1]**

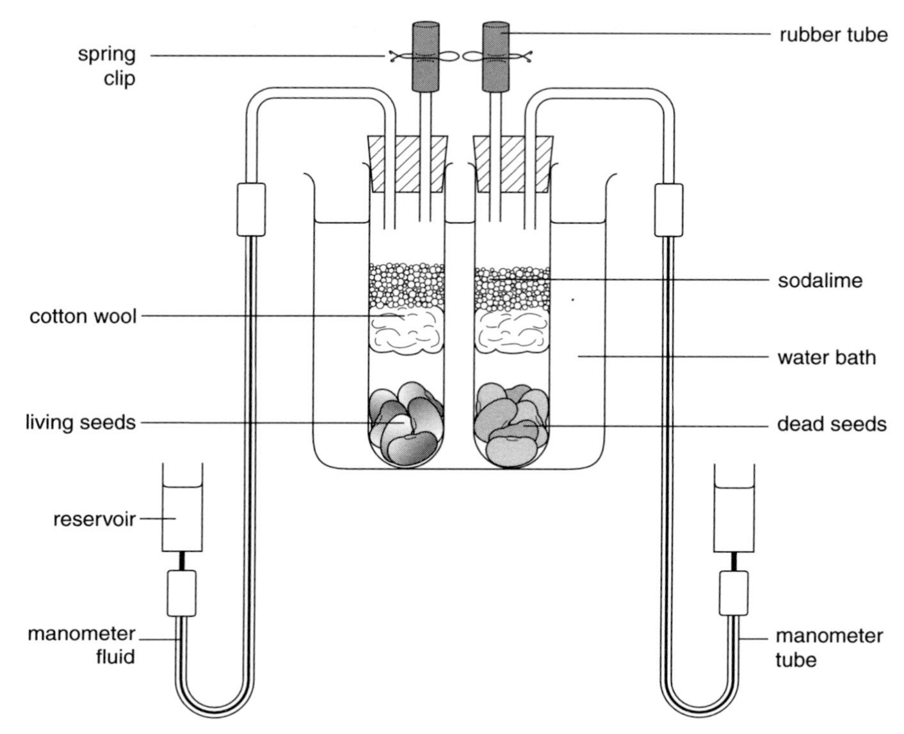

Figure 14.1.

14.8 Figure 14.2 shows the structure of a mitochondrion.

(a) Copy the figure and indicate on your diagram the location of the Krebs cycle and the electron transport chain. **[2]**

(b) What is the significance of the folded inner membrane? **[1]**

(c) What is the approximate length of a mitochondrion? **[1]**

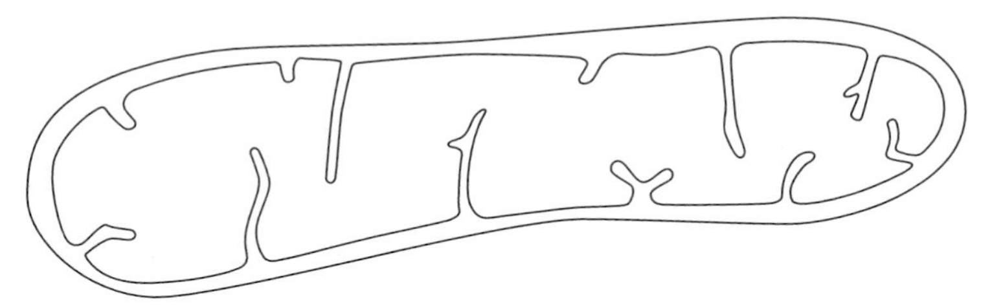

Figure 14.2

14.9 The following flow chart summarises the reactions that take place in glycolysis

$$\text{glucose} \rightarrow 2 \times \text{glyceraldehyde 3-phosphate} \rightarrow 2 \times \text{pyruvate}$$

(a) How many carbon atoms are there in glucose, glyceraldehyde 3-phosphate and pyruvate? **[3]**

(b) What is the net gain of ATP in glycolysis? **[1]**

(c) Why is ATP needed for glycolysis? **[1]**

(d) Hydrogen carriers are also involved in glycolysis. Name the hydrogen carrier and describe its role. **[2]**

(e) Where does glycolysis take place in the cell? **[1]**

(f) What happens to the pyruvate when **(i)** there is a plentiful supply of oxygen and **(ii)** there is no oxygen? **[2]**

14.10 **(a)** Where does the Krebs cycle take place in the mitochondrion? **[1]**

(b) What is the function of the Krebs cycle? **[2]**

(c) What happens to the products of the Krebs cycle? **[4]**

14.11 Describe the function of the electron transport chain. Where is it located in the mitochondrion? **[3]**

14.12 What does the term oxidative phosphorylation mean? **[2]**

14.13 In modern wine making, yeasts are added to the sugary grape juice and the mixture is allowed to ferment. First, the solution is aerated and then the air supply is cut off to allow anaerobic respiration. The temperature is kept low, so that fermentation is more complete and the yield of alcohol higher.

(a) Why are the grape juice and yeasts aerated at the start of the process? **[1]**

(b) One product of yeast fermentation is alcohol. What is the other product? **[1]**

(c) Suggest why fermentation is more complete at lower temperatures. **[2]**

(d) The alcoholic content of wine ranges from 7–14%. What prevents the alcoholic content from being higher? **[2]**

14.13 Tetrazolium chloride is a redox indicator which is colourless when oxidised and pink when reduced. This indicator acts as an artificial hydrogen acceptor and is readily reduced in the presence of dehydrogenase enzymes. The effect of temperature on the dehydrogenase activity of yeast cells was investigated. A 10 cm^3 sample of yeast suspension was mixed with 1 cm^3 tetrazolium chloride solution and placed in a water bath at 20 °C. The time taken for the solution to turn pink was noted. The investigation was repeated at temperatures from 0–50 °C. The results for the investigation are shown in the table.

Temperature of water bath / °C	*Time taken for solution to turn pink / minutes*	*Rate of reaction*
0	0 (no change)	
20	50	
25	36	
30	25	
35	19	
40	13	
45	20	
50	0 (no change)	

(a) Calculate the values for rate of reaction. **[3]**

(b) Plot a graph showing rate against temperature. **[4]**

(c) Describe the relationship between rate and temperature. **[2]**

(d) What is the optimum temperature for this reaction? **[1]**

(e) Calculate the Q_{10} for the yeast respiration between 20 and 30 °C. **[2]**

(f) Describe **two** precautions which should have been taken in this investigation. **[2]**

(g) What is the role of dehydrogenase enzymes in respiration? **[2]**

15 Photosynthesis

You should know that:

- green plants manufacture their own food by photosynthesis
- carbon dioxide + water → sugar + oxygen in the presence of light
- the green pigment, chlorophyll, found in chloroplasts, traps light energy and converts it to chemical energy
- oxygen is used in respiration of plants and animals
- the rate of photosynthesis depends on the light intensity, concentration of carbon dioxide and temperature
- the sugar is converted to starch for storage or used to synthesise new materials such as protein.

KEY FACTS

- Photosynthesis is the conversion of carbon dioxide and water into carbohydrates and oxygen in the presence of chlorophyll, using light energy.
- The site of photosynthesis is the chloroplast. Chloroplasts are found in leaf palisade, spongy mesophyll and guard cells.
- Photosynthesis has two stages – a light dependent stage and a light independent stage.
- During the light dependent stage, light excites an electron in the chlorophyll. The electron either passes along electron carriers back to the same chlorophyll in cyclic photophosphorylation, which results in the production of ATP, or passes to a different chlorophyll molecule in non-cyclic photophosphorylation, which produces ATP and NADPH. Light causes water to dissociate to form hydroxyl ions and hydrogen ions. The hydroxyl ions combine to form oxygen gas and more hydrogen ions. The ATP and NADPH are used in the next stage.
- During the light independent stage (Calvin cycle), carbon dioxide combines with ribulose biphosphate to form an unstable 6-carbon compound, which splits into two molecules of glycerate-3-phosphate. This compound is reduced using ATP and NADPH to form glyceraldehyde 3-phosphate. This can be used to form glucose.
- Chloroplasts contain stacks of membranes called thylacoids, these make up grana. The chlorophyll is found on these membranes and is the site of the light dependent stage. The light independent stage takes place in the stroma.
- The rate of photosynthesis is limited by the intensity and wavelength of light, the concentration of carbon dioxide and the temperature.

- The compensation point is when the rate of photosynthesis equals the rate of respiration and there is no net gain or loss of gases.

Questions for you to try

15.1 Name **two** types of plant cells in which chloroplasts can be found. **[2]**

15.2 Describe **three** ways in which the leaf is adapted for photosynthesis. **[3]**

15.3 Figure 15.1 shows the structure of a chloroplast.

(a) Identify the parts labelled A, B and C. **[3]**

(b) Where do the light dependent and light independent reactions take place? **[2]**

(c) This chloroplast has been magnified 10 000 times. Calculate the actual length of the chloroplast. **[2]**

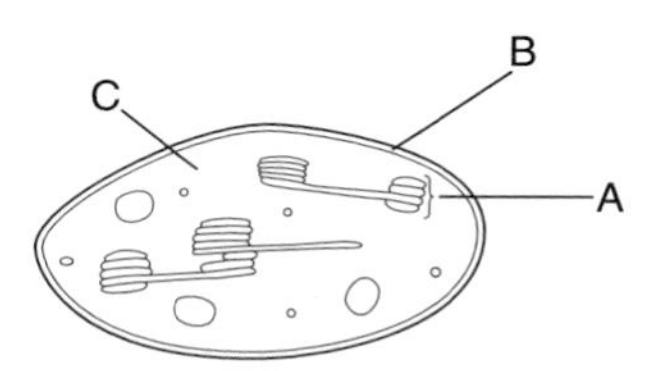

Figure 15.1

15.4 Compare and contrast the structure of a chloroplast with that of a mitochondrion. **[4]**

15.5 The apparatus in Figure 15.2 can be used to investigate the rate of photosynthesis.

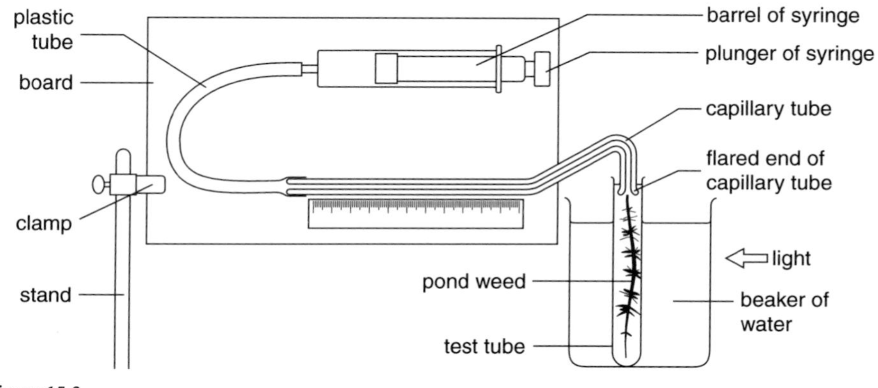

Figure 15.2

(a) Describe how you would set up this apparatus to measure the effect of light intensity on the rate of photosynthesis. **[3]**

(b) Describe **three** precautions you would need to take during this investigation. **[3]**

(c) Describe how you would modify the experiment to determine the effect on photosynthesis of **(i)** carbon dioxide concentration and **(ii)** the wavelength of light. **[4]**

15.6 Figure 15.3 summarises the movement of materials into and out of a chloroplast. Identify the substances moved, indicated by labels A–D. **[4]**

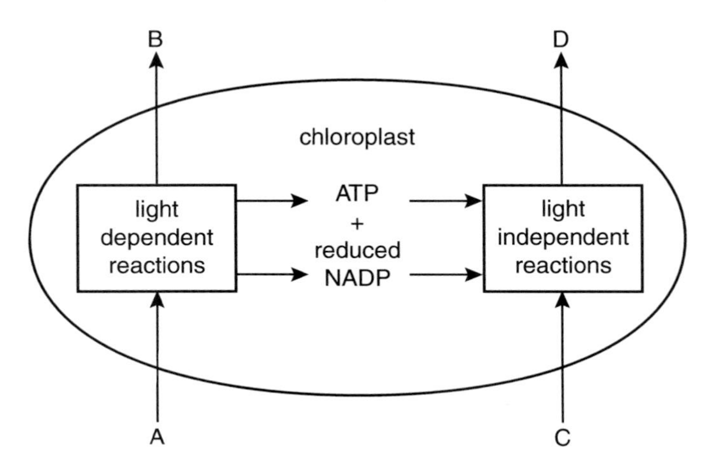

Figure 15.3

15.7 **(a)** What is **(i)** an absorption spectrum and **(ii)** an action spectrum? **[4]**

(iii) How are the two spectra linked? **[2]**

(b) Study the graph shown in Figure 15.4. What wavelengths of light are best absorbed by chlorophyll *a*? **[1]**

(c) How does the curve for chlorophyll *b* differ from that for chlorophyll *a*? **[2]**

(d) Why do plants have a number of photosynthetic pigments? **[2]**

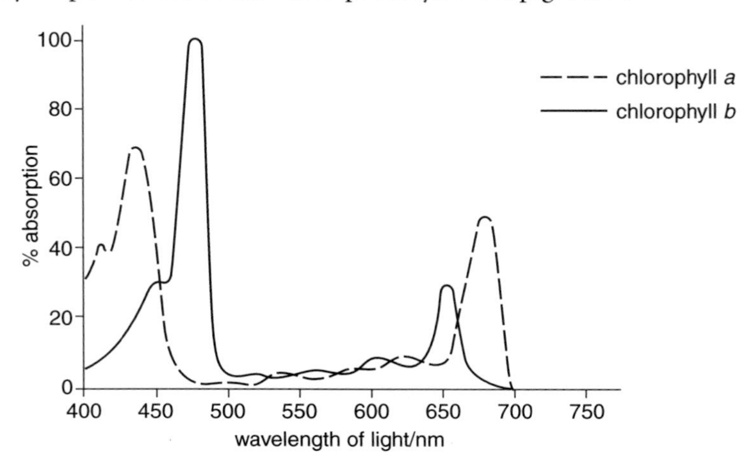

Figure 15.4

15.8 Figure 15.5 shows the sequence of events that takes place in the light dependent reactions.

(a) Identify the points labelled A and B. **[2]**

(b) What process is taking place at C? **[1]**

(c) What are the products of the light dependent reaction? (They are indicated by '?'s on the diagram.) **[3]**

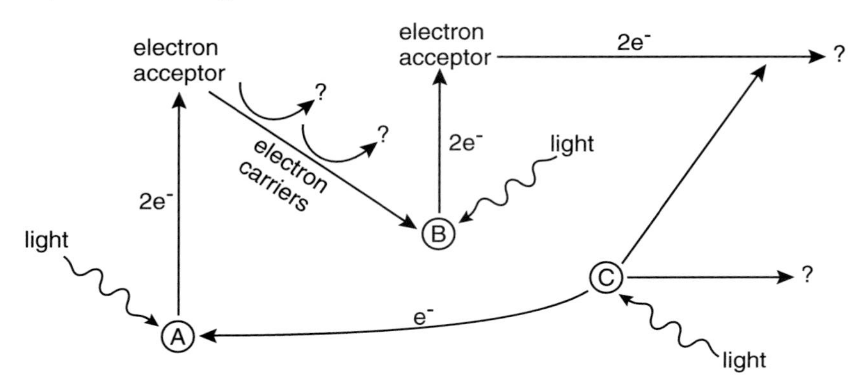

Figure 15.5

15.9 Only a small percentage of the sunlight falling on a pond is used by the pond weeds in photosynthesis. Can you suggest why this is so? **[2]**

15.10 An investigation was carried out into the effects of varying the concentration of sodium hydrogen carbonate on the photosynthetic rate of the pond weed, *Elodea.* The results are shown in Figure 15.6

(a) Describe the effect of increasing the concentration of sodium hydrogen carbonate on the rate of photosynthesis between light intensities of 0 and 200. **[2]**

(b) Why do the curves start to level off at a light intensity of 400? **[2]**

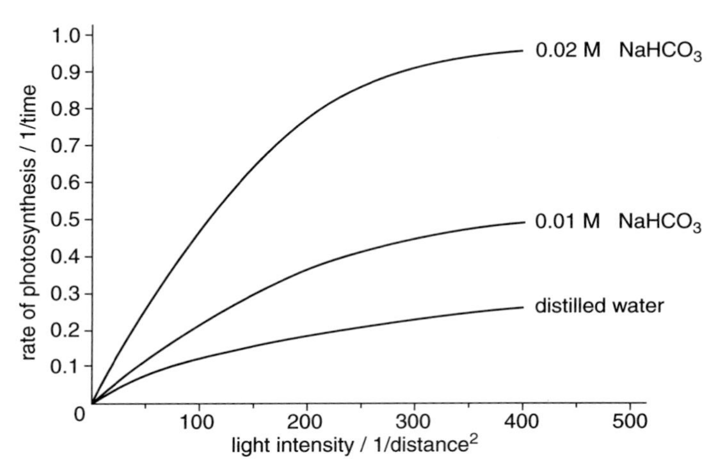

Figure 15.6

15.11 Two of the products of the light dependent reaction pass into the light independent reaction or Calvin cycle. This is summarised in Figure 15.7

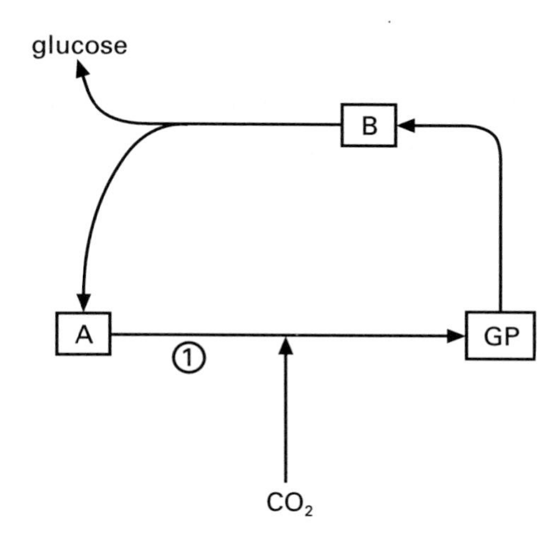

Figure 15.7

(a) Identify the substances labelled A and B. **[2]**

(b) What type of reaction is taking place at ①? **[1]**

(c) Describe how the two products from the light dependent reaction are used in this stage of photosynthesis. **[2]**

15.12 Figure 15.8 shows the effect of light intensity on carbon dioxide exchange by sun and shade leaves.

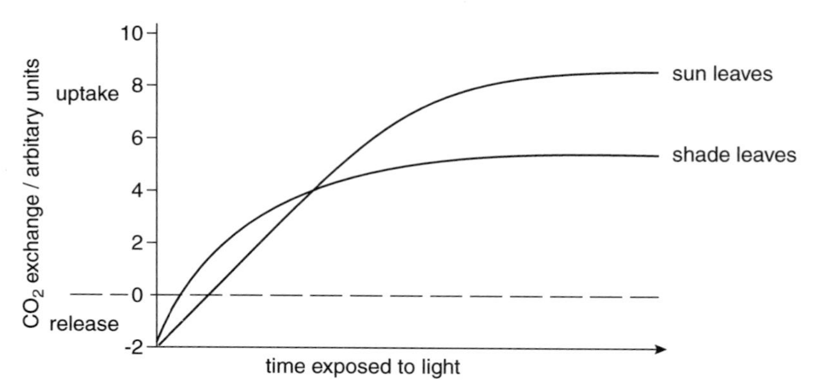

Figure 15.8

(a) What does the compensation point represent? **[3]**

(b) Why is the compensation point reached more quickly by the shade leaves than the sun leaves? **[2]**

(c) Compare the uptake of carbon dioxide of the two leaves. **[2]**

(d) How might the structure of a sun leaf differ from that of a shade leaf? **[2]**

(e) Suggest what would happen to the curve of the sun leaves if the plant was placed in a higher carbon dioxide environment? **[2]**

16 Ecology

You should know that:

- the environment is an organism's surroundings; the habitat is an organism's home; a population is the number of individuals of the same species living together in the same place; a community is all the animals and plants living in the same place
- plants are called producers and animals are consumers; herbivorous animals are primary consumers and carnivores are secondary and higher consumers
- energy is transferred along food chains
- food chains are inter-linked and form food webs
- a pyramid of numbers is a graphical representation of the numbers of plants and animals in a food chain; a pyramid of biomass shows the living mass of the plants and animals in a food chain and a pyramid of energy shows the energy in the plants and animals in a food chain
- decomposition is important in the recycling of nutrients, especially carbon and nitrogen
- both abiotic and biotic factors influence the size of populations.

KEY FACTS

- Energy enters ecosystems via photosynthesis.
- Feeding relationships are summarised in food chains and webs, pyramids of numbers and biomass.
- Energy is transferred along a food chain, via each trophic level.
- Energy flow is represented by a pyramid of energy and energy flow diagrams.
- Carbon enters the ecosystem via photosynthesis and most is returned by respiration and the burning of forests and fossil fuels.
- Micro-organisms have a key role in the cycling of nitrogen, carbon and phosphorus.
- The distribution of a species is determined by abiotic and biotic factors.
- Each species occupies a different niche within a habitat.
- The population of a species is the number of individuals living in a particular habitat.
- The size of a population is increased by birth and immigration and decreased by death and emigration.
- The carrying capacity is the maximum number of individuals of a particular species that can be supported by the environment.

- Environmental resistance will limit the size of a population, these limiting factors can be abiotic or biotic.
- The population of a species is affected by density dependent and density independent factors.
- There is competition between individuals of the same species (intra-specific) and between different species (inter-specific).

Questions for you to try

16.1 Fill in the gaps with the most appropriate word or words.

Ecology is the study of how living organisms interact with each other and with their _________. Ecosystems are made up of _________ and abiotic components. Two examples of abiotic components are __________ and __________. The _________ is the place where a particular organism lives. **[5]**

16.2 Define the terms 'community' and 'population'. **[4]**

16.3 Figure 16.1 shows a food web from the Antarctic Ocean.

(a) What are the producers in this food web? **[1]**

(b) Identify a primary, a secondary and a tertiary consumer. **[3]**

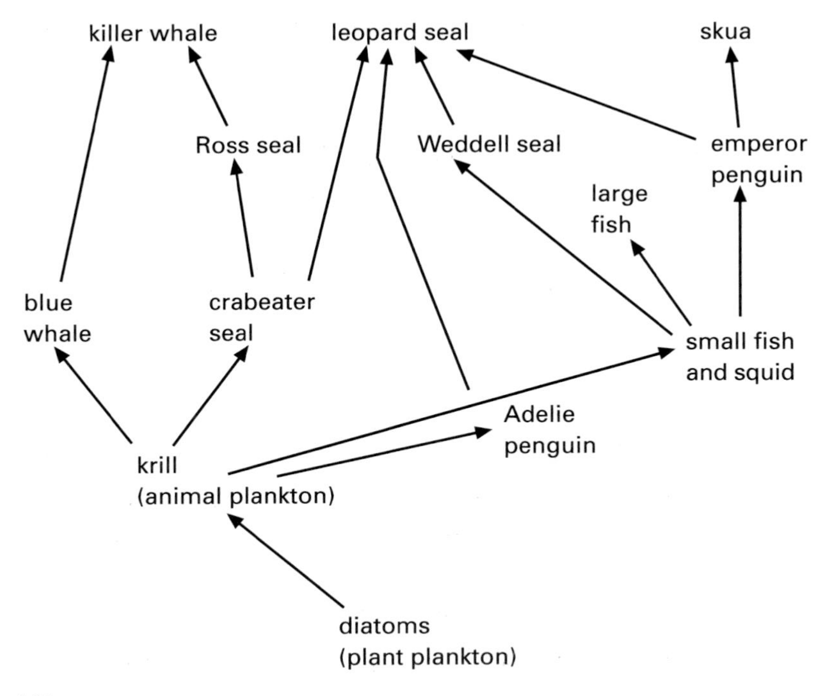

Figure 16.1

(c) Draw one complete food chain with at least four links. **[2]**

(d) Why is a food web more informative to ecologists than a food chain? **[2]**

(e) What would happen to this food chain if the numbers of small fish and squid were to decline? **[2]**

(f) Some of the food chains in the Antarctic Ocean are much longer than those seen in terrestrial habitats. Suggest a reason why the food chains are longer here. **[1]**

16.4 Figure 16.2 shows a pyramid of energy for a pond.

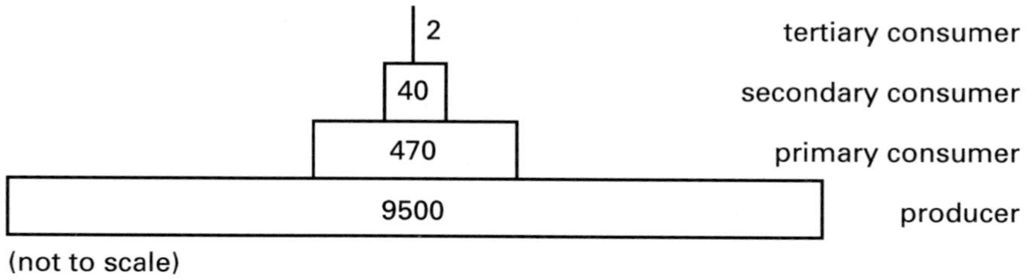

2 tertiary consumer
40 secondary consumer
470 primary consumer
9500 producer
(not to scale)

Figure 16.2

(a) There are no units given on this pyramid of energy. Suggest suitable units which could have been used to record energy. **[1]**

(b) Calculate the percentage energy transfer between the primary and secondary consumers. **[2]**

(c) Give **two** reasons why the percentage transfer of energy between trophic levels is low. **[2]**

(d) Why are pyramids of energy more informative than pyramids of biomass? **[2]**

16.5 Describe how you would sample a meadow in order to produce a pyramid of fresh biomass. **[6]**

16.6 The bottom of the pyramid of biomass for the English Channel is often inverted as shown in Figure 16.3. Suggest a reason for this inversion. **[2]**

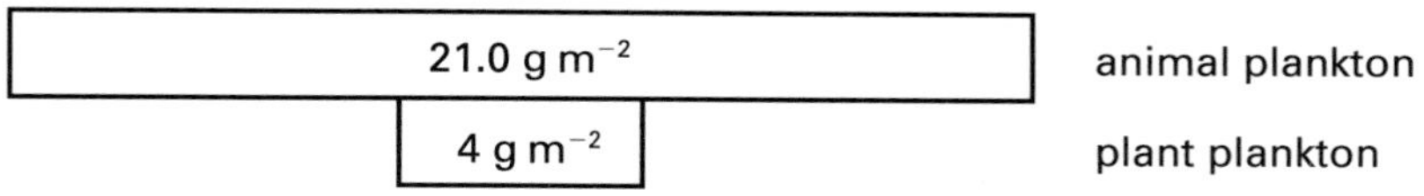

$21.0\ g\ m^{-2}$ animal plankton
$4\ g\ m^{-2}$ plant plankton

Figure 16.3

16.7 What is the difference between gross and net primary productivity? **[2]**

16.8 Distinguish between the following pairs of terms

(a) Density dependent factors and density independent factors. Give an example of each. **[4]**

(b) Inter- and intra-specific competition. **[2]**

16.9 Figures 16.4 shows the growth of a population of paramecium grown in the laboratory.

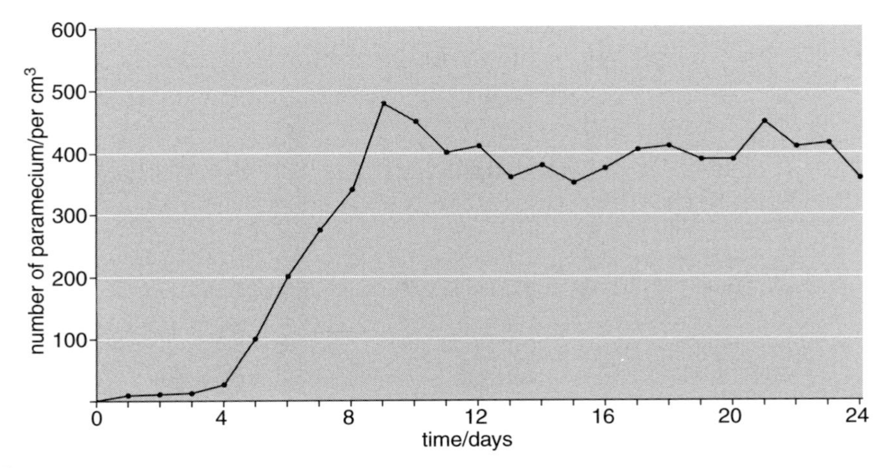

Figure 16.4

(a) Describe the population growth of the paramecium. **[3]**

(b) What is the approximate carrying capacity for the paramecium in this environment? **[1]**

(c) What factors control the carrying capacity? **[3]**

16.10 In 1955, the first collared doves appeared in Great Britain. Since that time the collared dove population has increased.

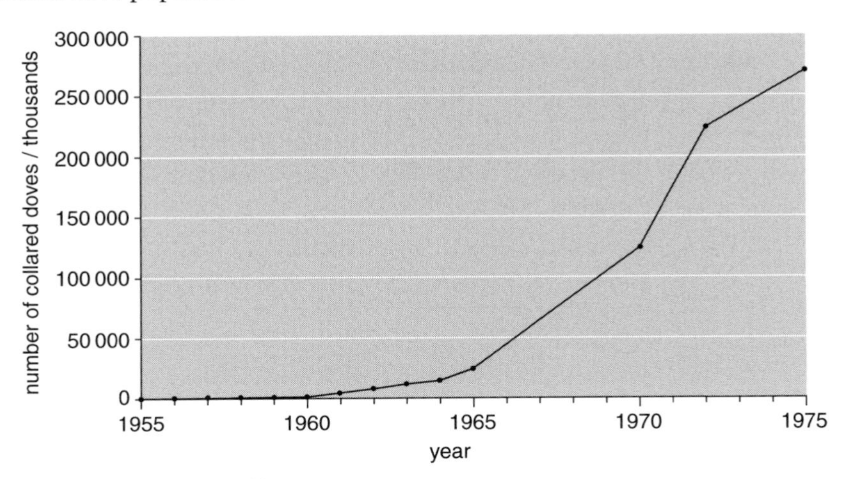

Figure 16.5 Data from Hengeveld 1988

(a) Why was there a slow increase between 1955 and 1963? **[2]**

(b) Explain the rapid increase in its population from 1965–1970. **[2]**

(c) Explain the slow down in the increase from 1970. **[2]**

16.11 The large cactus finch is found on the island of Genovesa in the Galapagos Islands. The Galapagos Islands have a highly variable climate, which affects the populations of the finches on the island. Figure 16.6 shows the age distributions of the large cactus finch in 1987.

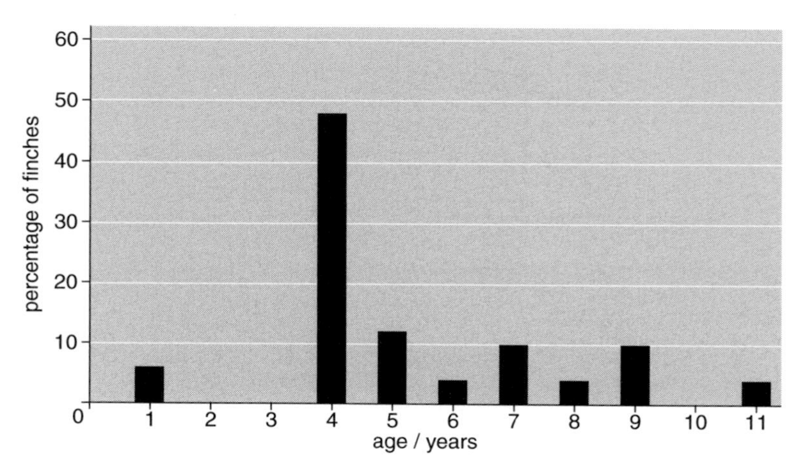

Figure 16.6 Data from Grant and Grant 1989

(a) Almost half the finches are four years old. What factors would have favoured this population boom? **[2]**

(b) What climatic conditions could have prevailed to account for no finches aged 2, 3 and 10 years? **[2]**

(c) What effect would the failure to breed in certain years have on the future of the species? **[2]**

16.12 An experiment was set up to investigate the effect of grazing by caddis flies on the algal population of a stream. The investigators placed small, unglazed ceramic tiles on the bottom of the stream and followed the colonisation of these tiles by the algae and caddis flies over a period of seven weeks. The results are shown in the table.

Weeks of colonisation	*Algal biomass/μg cm^{-2}*	*Number of caddis fly per tile*
0	0	0
1	0.65	25
2	0.81	30
3	0.68	137
4	0.27	103
5	0.13	100
6	0.35	95
7	0.75	50

(a) Plot these results on a graph. **[4]**

(b) Describe the similarities and differences between the two curves. **[3]**

(c) Explain the relationship between the two organisms. **[2]**

16.13 Explain the following terms:

(a) ecological succession, **[2]**

(b) climax vegetation, **[2]**

(c) biodiversity. **[2]**

16.14 Figure 16.7 shows the change in the number of bird species in secondary woodland that developed on abandoned fields over a period of 200 years.

(a) Describe the succession that would have taken place on the abandoned fields. **[3]**

(b) Describe the changes to the vegetation which allowed more bird species to colonise the habitat. **[2]**

(c) Explain why there was little change in the number of bird species during the last 100 years. **[2]**

(d) Suggest ways in which the woodland could be managed to increase the bird species further. **[2]**

(e) Explain what is meant by a 'deflected succession'. Give an example. **[3]**

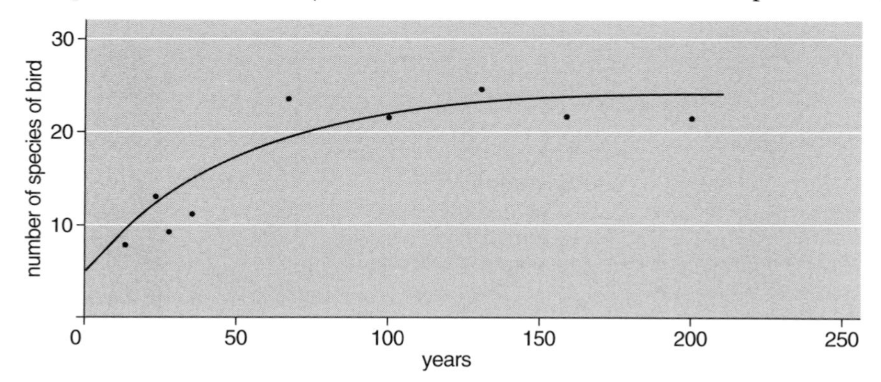

Figure 16.7

16.15 Explain the difference between

(a) line and belt transect, **[2]**

(b) frame and point quadrat, **[2]**

(c) random and systematic sampling. **[2]**

16.16 The capture–release–recapture method can be used to estimate the number of organisms in a population. In an investigation to estimate the population of woodlice living in a log pile, 100 woodlice were trapped and marked, then released. Two days later, a second sample of 90 woodlice were captured and 30 were found to be marked.

(a) Using the following formula, calculate the population of woodlice in the log pile. **[3]**

$$\frac{\text{number marked in second sample}}{\text{total caught in second sample}} = \frac{\text{number marked in whole population}}{\text{size of whole population}}$$

(b) Describe and explain **three** precautions that you would take when carrying out this experiment. **[3]**

16.17 Describe the method you would use to determine the abundance and distribution of plant species growing on a slope. **[5]**

16.18 Figure 16.8 shows the carbon cycle.

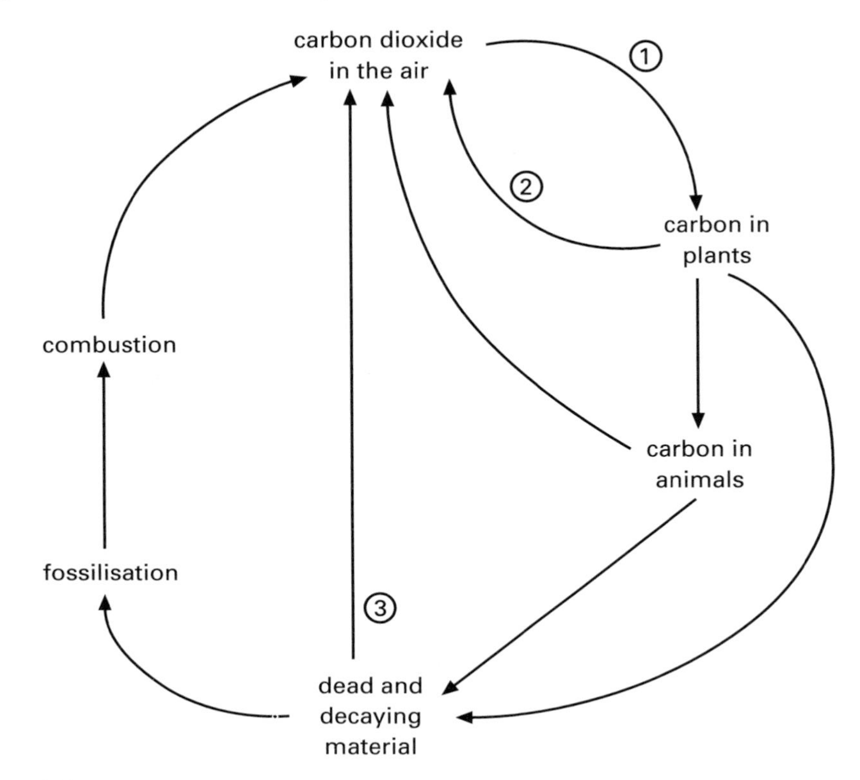

Figure 16.8

(a) Identify the processes labelled 1, 2 and 3. **[3]**

(b) Describe **two** ways by which carbon can be removed from the cycle for long periods of time. **[2]**

(c) Describe **two** activities of humans that are disrupting the natural carbon cycle. **[2]**

16.19 Figure 16.9 shows some of the stages in the nitrogen cycle. The forms of nitrogen are shown, but not the organisms that carry out the conversions.

(a) The bulk of the nitrogen exists as nitrogen gas in the atmosphere. What feature of nitrogen gas makes it unavailable to most organisms? **[1]**

(b) Name the organism that carries out stage 1. **[1]**

(c) Decomposers in the soil convert ammonia to ammonium ions. Name **two** types of decomposer. **[1]**

(d) Name the process that converts ammonium ions to nitrate ions in the soil. **[1]**

(e) Identify the specific bacteria that carry out stages 3 and 4. **[2]**

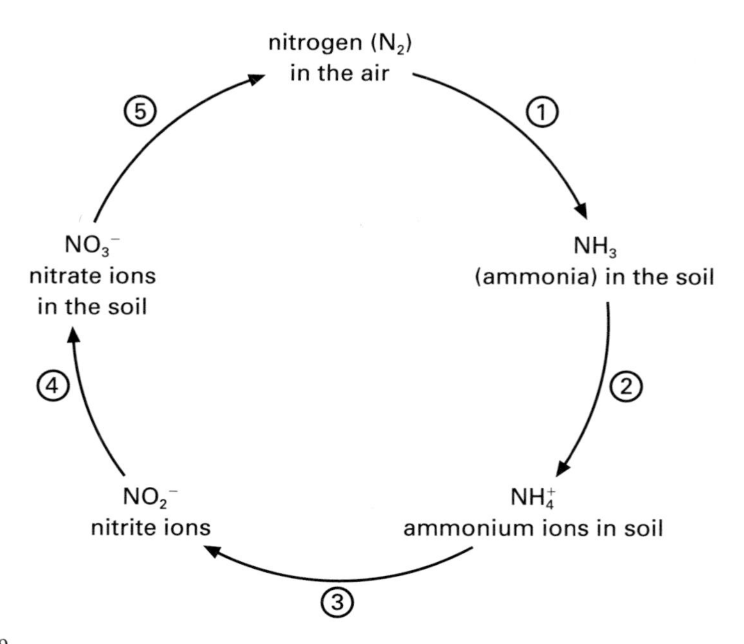

Figure 16.9

(f) In stage 5, some of the nitrate ion is denitrified back to nitrogen gas. Name a denitrifying bacterium and describe the conditions in which denitrification take place. **[3]**

(g) Beans and peas have root nodules that contain bacteria called *Rhizobium*. What is the role of the *Rhizobium* bacteria? Name the type of relationship that exists between the plant and *Rhizobium*. **[3]**

(h) How do plants take up the nitrate? **[2]**

(i) What is the role of nitrate in the plant? **[2]**

(j) How do animals acquire their nitrogen? **[1]**

(k) How is lightning involved in the nitrogen cycle? **[2]**

16.20 Write an account of the flow of energy through an ecosystem. **[8]**

17 Environmental issues

You should know that:

- human activity has polluted the air and water; sulphur dioxide causes acid rain, sewage and nitrates from fertilisers pollute water
- the level of human impact is related to the size of the human population, economic facts and industrial requirements.

KEY FACTS

- Renewable resources have an unlimited supply, whereas non-renewable resources have a finite supply as they can only be used once.
- As human population size increases, people create more waste and pollution and use up resources, such as fossil fuels and water.
- Fossil fuels release acidic gases when they burn in oxygen, creating acid rain that damages trees, aquatic habitats and buildings.
- Increases in greenhouse gases such as carbon dioxide and methane are contributing to global warming.
- CFCs are damaging the ozone layer, especially over Antarctica.
- Increased use of nitrogenous fertilisers and increased quantities of sewage are polluting water courses, causing eutrophication. This leads to algal blooms and loss of aquatic biodiversity.

Questions for you to try

17.1 What is meant by pollution? Give **two** examples of pollutants. **[2]**

17.2 Explain the term conservation. **[3]**

17.3 As the supply of fossil fuels runs out over the next 50 years, alternative sources of fuels such as wind and solar energy and biofuels will become increasingly important.

(a) What is a fossil fuel? **[2]**

(b) Why has the use of fossil fuels increased over the last 50 years? **[2]**

(c) In Brazil, petrol stations sell a fuel called gasohol. Gasohol is approximately 80% unleaded petroleum spirit and 20% ethanol. Describe how the ethanol can be produced from crops. Describe **two** of the benefits of using ethanol as a fuel. **[4]**

17.4 Figure 17.1 shows the increasing levels of nitrate in the River Lee, a tributary of the River Thames. Water entering this river drains off agricultural land.

(a) Suggest **two** possible sources of the nitrate in the water. **[2]**

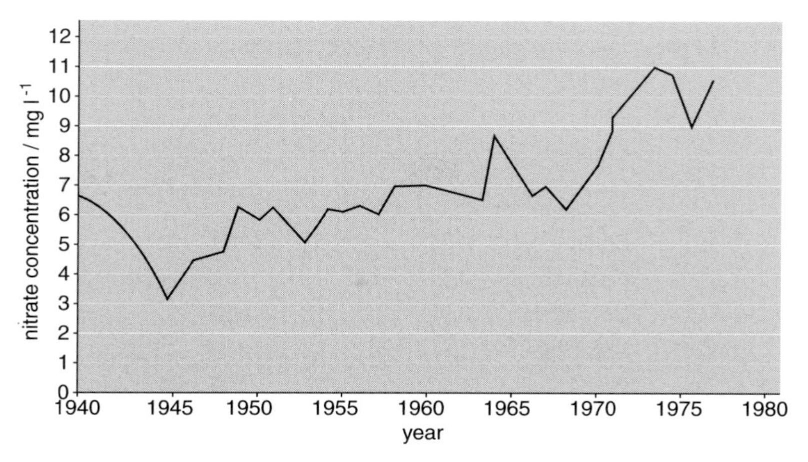

Figure 17.1

(b) How does the nitrate get into the river? **[2]**

(c) Why does the nitrate level fluctuate during the year? **[1]**

(d) Suggest why the nitrate levels in the water have increased since 1945. **[2]**

(e) Describe the effect of excess nitrate on the ecology of the river. **[4]**

17.5 **(a)** What is the role of the ozone in the atmosphere? **[2]**

(b) Name **two** chemicals that are destroying the ozone molecules. Give one use of each of these chemicals **[4]**

(c) What is being done by governments to protect the ozone layer from further damage? **[3]**

17.6 Explain how each of the following can be used to assess water pollution levels:

(a) biochemical oxygen demand, **[3]**

(b) indicator species, **[3]**

(c) species diversity index. **[3]**

17.7 Coppicing is a traditional method of woodland management.

(a) Describe the coppicing cycle. **[3]**

(b) What effect does coppicing have on the biodiversity of a woodland? **[2]**

(c) Suggest **two** uses for the wood produced by coppicing. **[2]**

17.8 The table shows the change in tropical forest cover in some countries of the world.

Country	*Area/million km²*	*Original forest cover/ thousand km²*	*Primary forest remaining in 1989/thousand km²*
Brazil	8.5	2860	1800
Ecuador	0.3	132	44
Venezuela	0.9	420	300
Indonesia	1.9	1220	530
Philippines	0.3	250	8
Zaire	2.3	1245	700
Madagascar	0.6	62	10

(a) Which country had the smallest percentage of their original forest cover remaining in 1989? Which had the highest? **[2]**

(b) Give **three** reasons for deforestation. **[3]**

(c) Describe **two** consequences of deforestation. **[2]**

(d) How does deforestation affect the global climate? **[2]**

17.9 **(a)** Explain the term desertification. **[2]**

(b) State **two** causes of desertification and for each explain how it increases the rate of desertification. **[4]**

(c) What can be done to reduce desertification? **[3]**

17.10 Acid rain has affected much of Western Europe

(a) Name **two** gases that contribute to acid rain. **[2]**

(b) What are the sources of these two gases? **[2]**

(c) Describe the effects of acid rain on **(i)** coniferous trees and **(ii)** lakes. **[6]**

(d) Scandinavian countries are sparsely populated, but more trees and lakes are damaged here, compared with other countries in Western Europe. Explain this high level of damage. **[2]**

(e) Describe **two** ways by which the production of acid rain can be reduced. **[2]**

17.11 **(a)** What is meant by the term bioaccumulation? **[2]**

(b) DDT is a persistent pesticide that was used during the 1960s. It is now banned in most countries. What does the term 'persistent pesticide' mean? **[1]**

(c) Why are the top predators in a food chain more likely to suffer from the toxic effects of the pesticide than organisms lower in the chain? **[2]**

(d) Modern pesticides tend to be selective. What does this mean? **[2]**

17.12 Figure 17.3 shows the increase in lichen cover outside the city of Belfast.

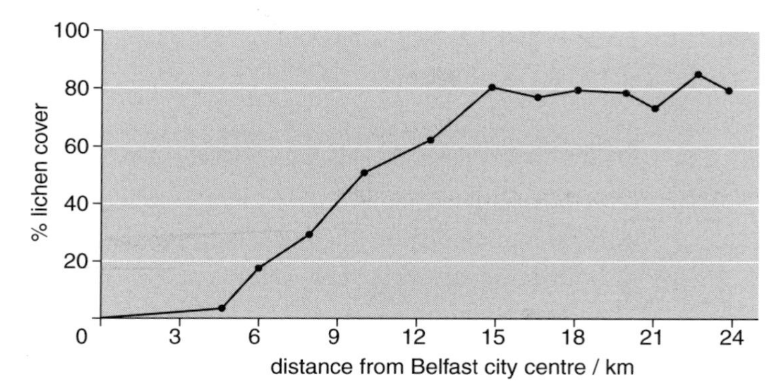

Figure 17.3 Data after Fenton, from Mellanby 1967

(a) Lichens are described as indicator species. What does this mean? **[2]**

(b) Describe the changes in lichen cover shown in the graph. **[2]**

(c) Suggest reasons for the change you have described. **[2]**

17.13 Figure 17.4 shows the annual herring catch in the North Sea.

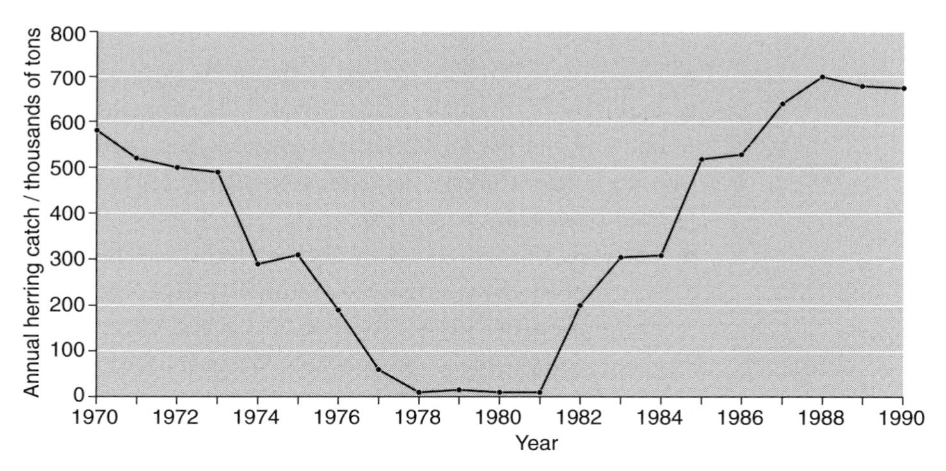

Figure 17.4

(a) Describe and explain the changes in the catch between 1970 and 1989. **[2]**

(b) Suggest what could have been done in 1970 to protect the herring stocks. **[3]**

18 Homeostasis

You should know that:

- homeostasis is the maintenance of a constant internal environment
- basal metabolic rate is a measure of the reactions going on in the body
- body temperature is kept within a narrow range by sweating, vasoconstriction and vasodilation
- excretory products include carbon dioxide and urea
- the kidney is responsible for excreting urea and excess salts as well as controlling water in the body, which involves the hormone ADH
- blood sugar levels are controlled by the hormones, insulin and glucagon; diabetics produce insufficient insulin, so need regular injections to control their sugar levels.

KEY FACTS

- Homeostatic mechanisms maintain the body in a state of equilibrium and give some independence from fluctuations in the external environmental conditions.
- Negative feedback mechanisms restore the systems to their original level.
- Endothermic animals regulate their body temperature within a narrow range and rely on their metabolism to generate heat energy. Ectotherms allow their body temperature to fluctuate more widely and rely on the environment for their body heat.
- The main thermoregulatory organ in humans is the skin. When the temperature rises, the arterio-ventricular shunt closes and more blood flows nearer the skin's surface so more heat is lost, sweat is produced from sweat glands and the metabolic rate falls. When the temperature falls, the shunt opens so blood flows deeper in the skin, hairs are raised to trap air and an increased metabolic rate and shivering generates heat.
- Control mechanisms are initiated by temperature receptors in the hypothalamus.
- Blood glucose levels are controlled by the pancreatic hormones insulin and glucagon. They travel in the blood to their target organ, the liver. Most body cells have insulin receptors on their cell membrane and once bound, insulin lowers blood glucose levels by increasing the rate of respiration, increasing the conversion of glucose to the storage carbohydrate glycogen and to fat, and increasing glucose uptake in muscle cells. Diabetics lack sufficient insulin and have to control their diets and inject themselves with insulin.
- Excretion is the disposal of material which would become toxic if allowed to

remain in the body. Carbon dioxide and urea are the main excretory products in mammals, fish produce ammonia and insects produce solid uric acid.

- Excess proteins are deaminated in the liver, producing urea which is excreted by the kidney.
- The kidneys are osmoregulatory and excretory. Ultrafiltration takes place in the Bowman's capsule, selective reabsorption takes place in the first convoluted tubule. Glucose, amino acids, hormones, sodium ions and some urea are actively reabsorbed into the capillaries, while water and chloride ions follow passively. The filtrate is concentrated by the counter-current mechanism of the loop of Henle. The second convoluted tubule regulates the level of sodium and controls the blood pH. Osmoreceptors in the hypothalamus detect changes in blood concentration and send impulses to the pituitary to release ADH. This hormone increases the permeability of the second convoluted tubule and collecting duct, allowing water to pass out and the filtrate to become more concentrated.
- Desert animals conserve water by having extra long loops of Henle to produce a more concentrated urine, they make use of water in food stuffs and can use metabolic water released from respiration.

Questions for you to try

18.1 Define the term 'homeostasis'. Give one example of a homeostatic mechanism. **[3]**

18.2 What does the phrase 'negative feedback' mean? How does negative feedback differ from positive feedback? **[3]**

18.3 **(a)** Define the terms 'excretion' and 'osmoregulation'. **[3]**
(b) Name **three** excretory organs in a human and list their excretory products. **[3]**

18.4 Figure 18.1 shows the structure of the kidney. Identify the structures labelled A–F. **[6]**

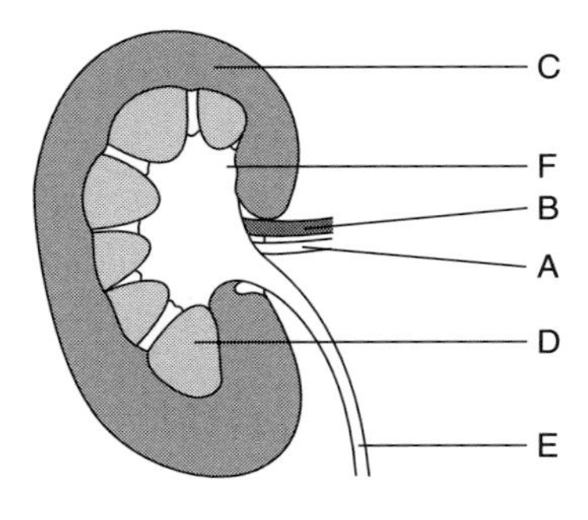

Figure 18.1

18.5 Where does the process of deamination take place? **[1]**
How is an amino acid converted to urea? **[2]**

18.6 What is an osmoreceptor? Where are they located in a human? **[3]**

18.7 What is the role of the pituitary gland in osmoregulation? **[2]**

18.8 Define the phrase 'core body temperature'. **[2]**

18.9 State **three** physical ways in which heat can be lost from the body. **[3]**

18.10 Read through the passage below and then fill in the gaps with the most appropriate word, words or numbers:

> The normal body temperature of a human is _________ °C. When the blood temperature rises, receptors in the _________ send __________ to the heat loss centre which initiates mechanisms to lose heat from the body. There is an _________ in the rate of sweating which loses heat by __________. Superficial blood vessels in the skin undergo _________ which __________ the amount of heat lost by convection and ___________ from the skin's surface. **[8]**

18.11 Figure 18.2 is a photomicrograph of kidney tissue

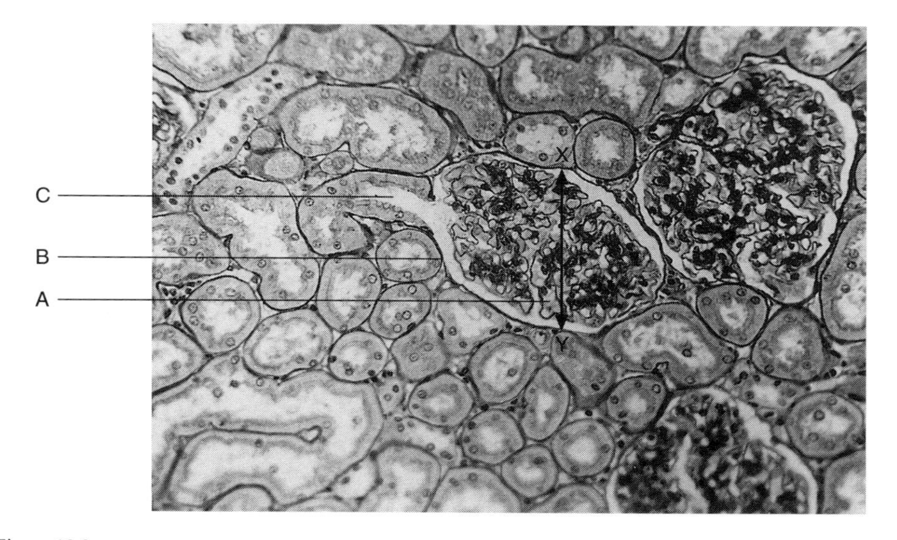

Figure 18.2

(a) Identify the structures labelled A, B and C. **[3]**

(b) The photo has been magnified by 300 times. Measure the diameter of structure A along the point marked X – Y and calculate the actual diameter. Show all of your working. **[2]**

18.12 When the body temperature falls, a number of mechanisms bring the body temperature back up to normal. Describe these mechanisms. **[2]**

18.13 Name the condition in which the body's temperature falls to below 32 °C. **[1]**

18.14 Read through the passage below and then fill in the gaps with the most appropriate word or words:

Blood glucose levels rise after a meal. This causes the _______ cells in the islets of _______ to release _______ . The glucose levels fall as a result of glucose being converted to _______ in the liver and _______ . If blood glucose levels fall below normal _______ is released from the _______ cells in the pancreas. **[7]**

18.15 The nitrogenous waste product of fish is ammonia, while insects produce uric acid. How is the nitrogenous waste product of these organisms linked to their environment? **[3]**

18.16 What is hypoglycaemia? **[1]**

18.17 Figure 18.3 shows the change in blood glucose levels in a normal person and a diabetic at hourly intervals after a meal

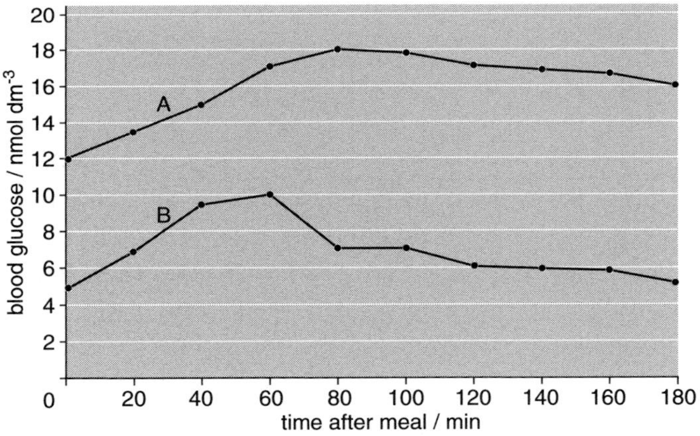

Figure 18.3

(a) Compare the two curves over the period of the experiment. **[3]**
(b) Which curve represents the diabetic? Explain your answer. **[3]**
(c) What causes diabetes? **[2]**
(d) How can diabetes be treated? **[2]**

18.18 What are the advantages of human insulin produced by genetic engineering compared with that extracted from the bodies of cattle? **[3]**

18.19 **(a)** Distinguish between the terms ectothermy and endothermy. **[3]**
Figure 18.4 shows the change in body temperature with environmental temperature for two animals.
(b) Which of the curves represents an ectotherm? Explain your answer. **[3]**
(c) What restrictions does the physiology of ectotherms place on their geographical distribution? **[2]**

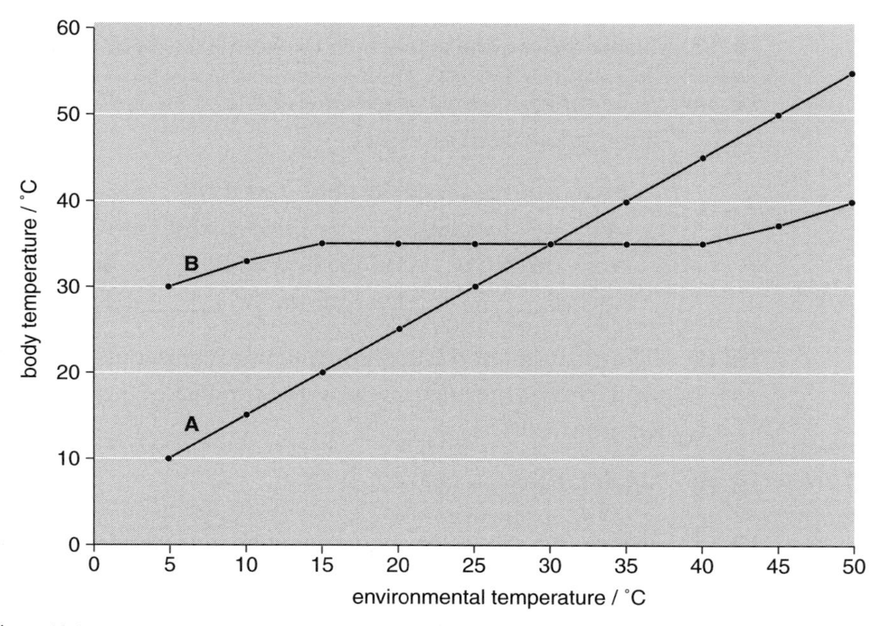

Figure 18.4

18.20 Figure 18.5 shows the structure of the glomerulus and the Bowman's capsule in the kidney.

(a) What is ultrafiltration? **[2]**

(b) What is the significance of the difference in diameter between the afferent and efferent arteriole? **[2]**

(c) What materials pass into the Bowman's capsule? **[2]**

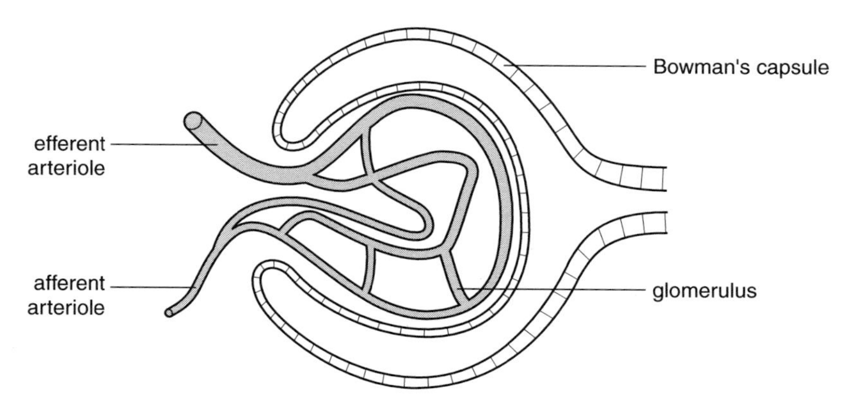

Figure 18.5

18.21 The loop of Henle and the surrounding capillary network form a counter-current mechanism. Describe how this mechanism works. **[4]**

18.22 **(a)** Why do many desert animals have a long loop of Henle? **[1]**

(b) Desert animals often rely on metabolic water. Explain this term. **[2]**

(c) How can desert animals modify their behaviour to conserve water? **[2]**

18.23 Anti-diuretic hormone (ADH) is involved in the regulation of water loss.

(a) Where is it produced? **[1]**

(b) When is it produced? **[1]**

(c) How does it affect the nephron? **[2]**

18.24 The table below shows the quantities of substances which are filtered, reabsorbed or excreted in the kidney

Substance	*Quantity filtered into nephron each day*	*Quantity reabsorbed per day*	*Quantity excreted per day*	*% of filtered which is reabsorbed*
water	180 litres	178.5	1.5	
glucose	800 mEq	799.5	0.5	
urea	56 g	28	28	
sodium ions	25 200 mEq	25 050	150	
chloride ions	18 000 mEq	17 850	150	
potassium ions	720 mEq	620	100	

(a) Complete the last column by calculating the percentage of the filtered quantity which is reabsorbed. **[6]**

(b) Why is nearly all of the glucose reabsorbed? **[2]**

(c) What disease is characterised by quantities of glucose in the urine? **[1]**

(d) Urea is a nitrogenous waste product. Why is half of it reabsorbed? **[2]**

18.25 Figure 18.6 shows the detailed structure of a cell of the proximal/first convoluted tubule and the adjacent capillary.

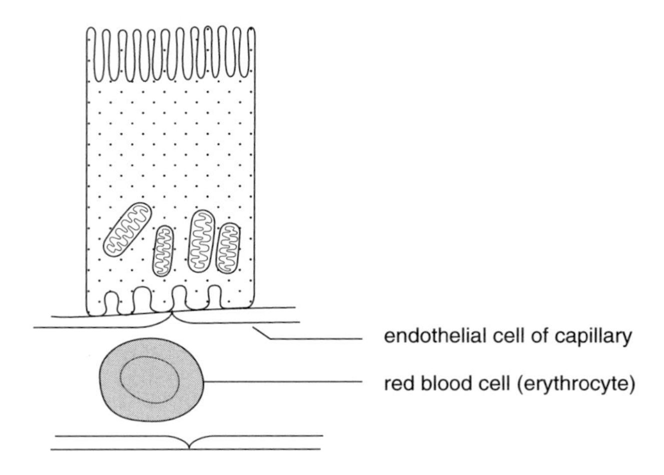

Figure 18.6

(a) How is the structure of this cell adapted to absorb materials from the tubule? **[2]**

(b) Describe the way in which glucose and water are moved from the filtrate into the capillary. **[2]**

18.26 Draw a flow chart to show how ADH operates by negative feedback. **[4]**

19 Nervous and chemical co-ordination

You should know that:

- sensory nerves carry impulses from the receptors to the central nervous system; motor nerves carry impulses from the central nervous system to the effectors, such as muscles
- nerves are made up of many neurones, bundled together like wires in a cable
- a nerve impulse is an electrical signal that passes down the axon of a neurone
- reflexes are rapid reactions that do not involve the brain
- the central nervous system comprises the brain and spinal cord; the different parts of the brain have different functions
- the eye is a sense organ that converts light energy into the electrical energy of an impulse; the brain interprets the impulses to create an image and the eye focuses the image onto the retina using the cornea and lens
- endocrine glands release hormones into the blood stream and they travel to target organs where they have their effect
- hormones are slow acting; insulin and glucagon control blood sugar levels, oestrogen, progesterone and testosterone are a few of the hormones that control human reproduction
- plant growth is controlled by hormones such as auxin.

KEY FACTS

- A nervous system is made up of neurones which carry impulses from receptor cells, giving information about the external environment. Impulses travel to effectors, such as muscles, to bring about a response.
- There are three types of neurone; sensory, motor and relay (also called connector). They all have a cell body containing a nucleus, dendrites which connect with other neurones and an axon which carries the impulse away from the cell body. In some neurones, the axon is covered by a myelin sheath which speeds up the transmission of the impulse.
- A nerve impulse is a minute electrical event produced by charge differences across the membrane of the axon. While not conducting an impulse, the membrane has a negative resting potential due to differential permeability. This allows the build up of positive charged ions, such as sodium, on the outside, so the outside of the neurone is positive and the inside is negative. A stimulus causes the permeability to change and the sodium ions flood in, reversing the potential. This depolarisation lasts a few milliseconds and is called the action potential. During the refractory, or recovery, period the sodium is pumped out and the resting potential is regained.

- A synapse is the point where two neurones meet. The arrival of an impulse causes a neurotransmitter to be released into the gap between the two neurones and this brings about depolarisation in the second neurone.
- In a reflex arc, a stimulus sets up an impulse in the sensory neurone which travels along the sensory neurone and synapses with a relay neurone, which in turn synapses with a motor neurone in the spinal cord. An impulse travels down the motor neurone to an effector.
- Sense organs, such as the eye, contain receptors which are sensitive to particular stimuli. The photoreceptors in the retina are called rods and cones. Rods contain rhodopsin which is bleached by light and gives black and white vision. Cones are responsible for colour vision. There are three forms of the pigment iodopsin which is found in cones, each sensitive to a colour – red, blue and green.
- The autonomic nervous system is subdivided into sympathetic (generally excitory) and parasympathetic (generally inhibitory) systems.
- The central nervous system consists of the brain and spinal cord. The brain has specialised areas: cerebrum for voluntary behaviour and conscious thoughts, cerebellum for co-ordinating movement and balance, medulla for reflexes and autonomic control.
- Chemical control has a long lasting effect and is brought about by hormones which are secreted by endocrine glands directly into the blood. Release of hormones is controlled by negative feedback. Hormones bind to receptor sites on their receptor cells. They may affect membrane permeability or cause the formation of a messenger molecule in the cytoplasm.
- Plant growth substances control plant growth. Auxin is responsible for phototropism, cell elongation, apical dominance; gibberellins cause stem elongation and branching; cytokinins stimulate cell division in the presence of auxin. These three substances are growth promoters. Abscisic acid brings about dormancy and leaf fall. Ethene stimulates the ripening of fruit. The complementary action of two or more growth substances is called synergism, while the inhibition of one substance by another is called antagonism.

Questions for you to try

19.1 **(a)** What is a hormone? **[2]**

(b) Give **one** example of a hormone, and state where it is made and where it has its effect. **[3]**

19.2 What is meant by the term central nervous system? **[2]**

19.3 **(a)** What triggers the release of a hormone? **[2]**

(b) Give an example. **[1]**

19.4 Read through the passage below and then fill in the gaps with the most appropriate word or words.

Chemical control in animals is brought about by _______. These chemicals are usually either protein or _______. They are produced by _______ which release them into the ________ where they travel around the body until they reach the _______ organ. This is where they bring about a response. **[5]**

19.5 How do hormones bring about changes within cells? **[2]**

19.6 Produce a table which compares and contrasts the features of nervous control with those of chemical control. **[5]**

19.7 The pituitary gland is often described as the 'master gland'. Explain this phrase. **[2]**

19.8 Figure 19.1 shows a motor neurone.

(a) Identify the structures labelled A–F. **[6]**

(b) What is the role of a motor neurone? **[2]**

(c) What are the names of the other two types of neurone? **[2]**

(d) What is the role of the structure labelled D? **[2]**

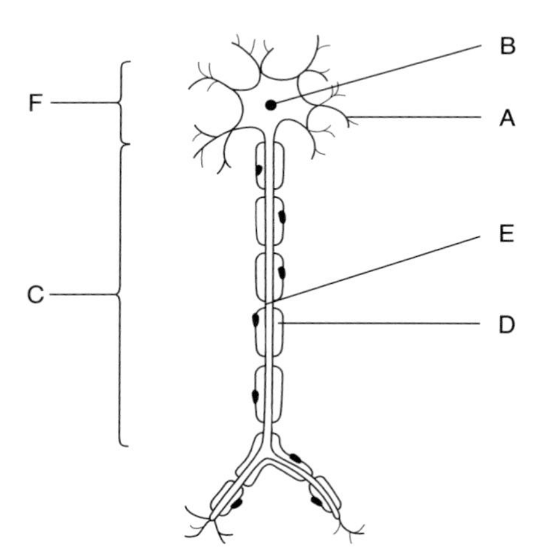

Figure 19.1

19.9 Figure 19.2 shows a cross section through a spinal cord of a mammal.

(a) Identify the structures labelled A–E. **[5]**

(b) Copy the figure on the left hand side and draw and label three neurones (simple lines only) to show a spinal reflex arc. **[3]**

(c) Indicate by means of arrows, the direction in which the impulse will travel along the three neurones. **[1]**

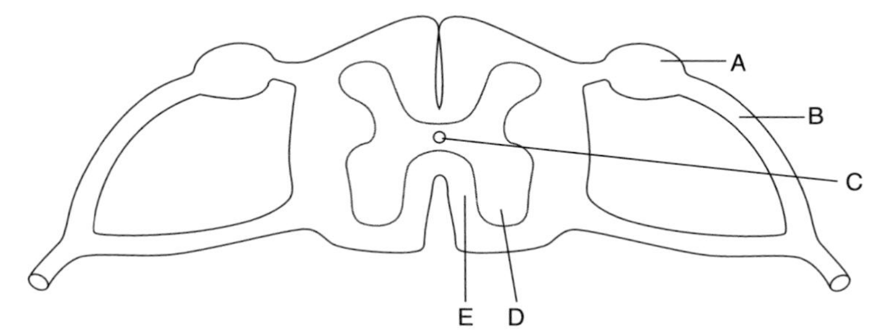

Figure 19.2

19.10 All hormones work on a similar principle. The flow chart below shows the sequence of events leading to the production of thyroxine.

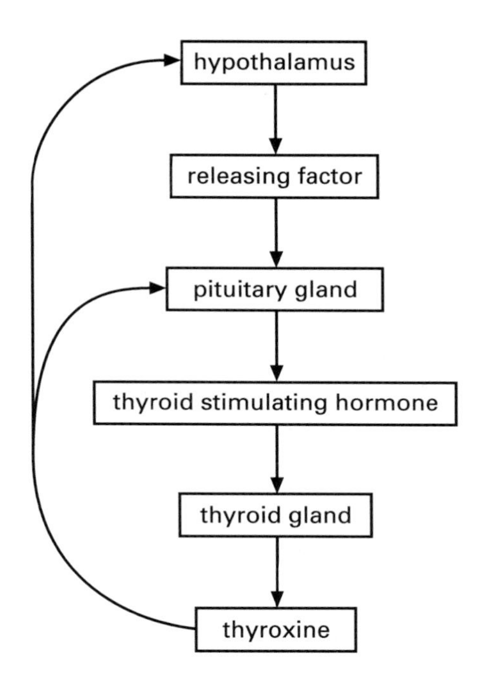

(a) How does the level of thyroxine affect the action of the hypothalamus and pituitary gland? **[3]**

(b) What is the name of this control mechanism? **[1]**

(c) Thyroxine affects both the hypothalamus and the pituitary gland. Is there any advantage in affecting both, rather than one or the other? **[1]**

(d) How does the thyroid stimulating hormone reach the thyroid gland? **[1]**

19.11 What is the role of plant growth substances in plants? **[2]**

19.12 Copy and complete the following table. **[10]**

Plant growth substance	*Site of synthesis*	*Effect in plant*
auxin		
gibberellin		
cytokinin		
abscisic acid		
ethene		

19.13 Auxins have a number of commercial applications. Describe **two**. **[2]**

19.14 Describe the role of gibberellins in seed germination. **[3]**

19.15 Figure 19.3 shows the structure of a human brain. Identify the structures A–D and for each structure give **one** function. **[8]**

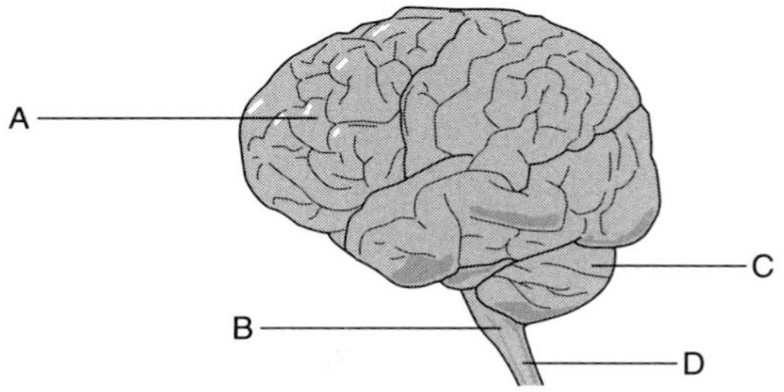

Figure 19.3

19.16 Draw a simplified diagram of a synapse. Show the following structures on your diagram: pre-synaptic membrane, post-synaptic membrane, synaptic cleft, mitochondria and vesicles. **[8]**

19.17 **(a)** What is the difference in meaning between the terms 'antagonistic' and 'synergistic' when referring to plant growth substances? **[2]**

(b) Which **two** plant growth substances act antagonistically and which act synergistically? **[2]**

19.18 Describe the sequence of events which take place when a nerve impulse arrives at a synapse up to the point when a new nerve impulse is generated in the post synaptic neurone. **[6]**

19.19 Figure 19.4 shows the change in membrane potential during the passage of a nerve impulse.

(a) What is the resting potential of this neurone? **[1]**

(b) How is the resting potential maintained in the neurone? **[3]**

(c) Explain how ion movements bring about the change in membrane potential between points A and B on the graph. **[2]**

(d) How is the resting potential restored? **[3]**

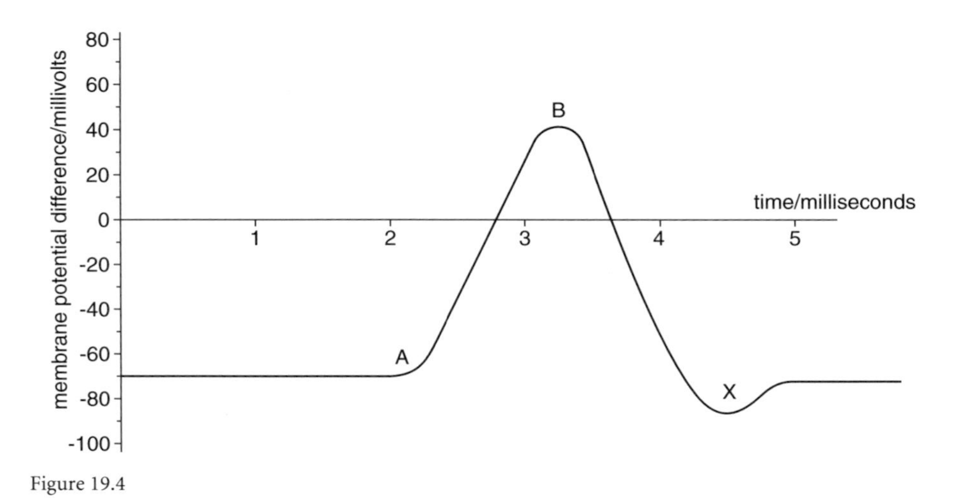

Figure 19.4

(e) What is the refractory period? **[1]**

(f) How does the length of the refractory period limit the number of impulses which can pass along the axon? **[1]**

(g) Account for the blip at point X on the graph. **[2]**

19.20 Read the passage below and then fill in the gaps with the most appropriate word or words.

Light enters the eye through a transparent layer called the ______ . It then passes through the lens. The lens can change shape to alter its focal length. This is called ________. In order to focus on a near object, the lens becomes _______ and _______. This is bought about by the contraction of the ________ which reduces the tension in the ________ . The lens bends the light so that it is brought to a focus on the _______ . Here, there are two types of photoreceptor cell. The ________ are responsible for seeing in low light intensities and at night. They contain the pigment _______ . The _______ are responsible for colour vision. They are found packed together at a point called the ________ . The nerve impulses generated by the photoreceptor cells are carried by the _______ nerve to the brain. **[12]**

19.21 How does the iris control the amount of light entering the eye? **[3]**

19.22 Describe the trichromatic theory of colour vision. **[3]**

19.23 A student investigated the growth of lateral shoots in pea plants. Two groups of 10 plants were selected. One group had their apical buds removed, while the other group, the control, were left intact. At various time intervals the total length of the lateral shoots were measured and the mean lateral shoot length calculated. The results are shown in the table.

Time / days	*Mean length of lateral shoot / mm*	
	Group A – apical bud removed	*Group B – apical bud intact*
2	4	4
4	12	4
6	40	4
8	80	4
10	120	4
12	160	4

(a) Plot these results on a graph. **[5]**

(b) Describe the curve for the group A plants. **[2]**

(c) How does the apical bud control the growth of lateral shoots? **[2]**

(d) Describe **two** precautions you would take while carrying out this investigation. **[2]**

20 Synoptic questions

All candidates taking A2 level biology will have to answer synoptic questions. Synoptic assessment makes up 20% of the A2 level marks.

Synoptic questions bring together principles and concepts from the different areas of the specification and apply them in a particular context. You will be expected to be able to express your ideas clearly and logically and to use the appropriate specialist vocabulary.

Synoptic questions include the reading and interpretation of a passage of text; interpretation of data, diagrams, flow charts and photographs; evaluating data; performing calculations; essays. The questions pull together topics studied at AS level as well as those at A2. For example, you may have to use the information you learn about proteins and enzymes at AS level to answer a question based on metabolic pathways.

Structured questions

20.1 Sickle cell anaemia is a genetic disease in which normal haemoglobin (HbA) is replaced by an abnormal haemoglobin (HbS). HbS has a lower affinity for oxygen and at low oxygen concentrations HbS polymerises and becomes insoluble. This alters the shape of the red blood cells, causing them to become sickle-shaped. The life span of these cells is greatly reduced from 120 days to just a few days, causing severe anaemia. In addition, the individual may suffer from extremely painful periods when the sickle cells block small blood vessels. As a result, the sufferers have a much reduced life expectancy. The alleles of the gene that causes sickle cell anaemia are codominant. A heterozygote experiences mild symptoms of the disease.

(a) What type of mutation causes sickle cell anaemia? **[1]**

(b) What effect does this mutation have on the genetic code? **[2]**

(c) Explain the meaning of the phrase 'lower affinity for oxygen'. **[2]**

(d) What does the term 'codominant' mean? **[2]**

Malaria is a disease caused by the parasite, *Plasmodium falciparum*, which is carried by the female mosquito. The parasites enter the bloodstream while the mosquito takes a blood meal from a human. They invade red blood cells where they multiply in number and use up the oxygen carried on the haemoglobin. The low oxygen concentration in the blood of a person with HbS kills the parasites.

Figure 20.1 shows the distribution of malaria and sickle cell anaemia in Africa

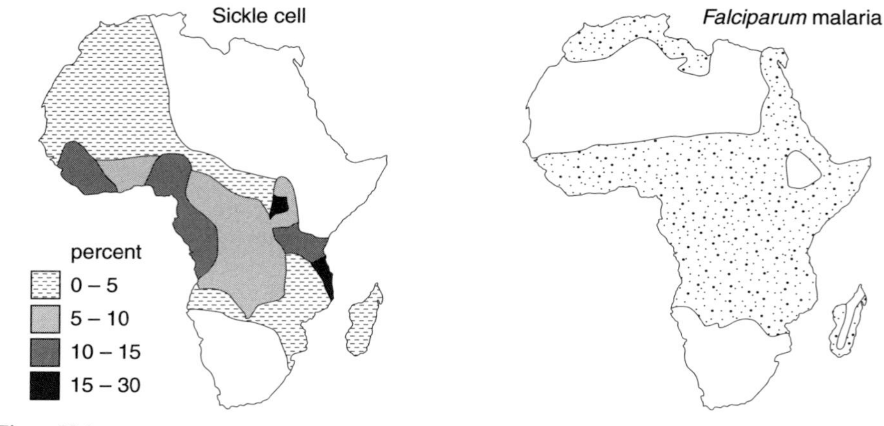

Figure 20.1

(e) Compare the distribution of malaria in Africa to that of sickle cell anaemia. **[2]**

(f) Suggest why there is no incidence of sickle cell anaemia in Southern Africa. **[2]**

(g) Explain why heterozygotes carrying the HbS gene are at an advantage compared with people who are homozygous for the condition. **[2]**

(h) Many Africans taken to North America as slaves carried the gene for sickle cell anaemia. Sickle cell anaemia is still present in the Afro-American population of the USA. There is no malaria in the USA. Suggest how the frequency of the HbS gene will change over the next 100 years. **[2]**

20.2 Fields of yellow oil seed rape and blue linseed are a common sight in early summer. Both these crops are grown for oil. As well as producing edible oil for the food industry, oil crops are a source of non-edible products for the pharmaceutical and cosmetic industries. They can also be used as a fuel.

The oils accumulate in the seeds and fruits of the plants. Figure 20.2 shows how sucrose from photosynthesis can be used to make storage oil and membrane lipids in plants.

(a) Why is oil stored in the seed? **[2]**

(b) How is sucrose moved around the plant? **[2]**

(c) Where in a cell does glycolysis take place? **[1]**

(d) Figure 20.2 shows how acetyl CoA is a source of membrane fatty acids and how it can be modified to make other fatty acids. Describe one other fate of acetyl CoA. **[2]**

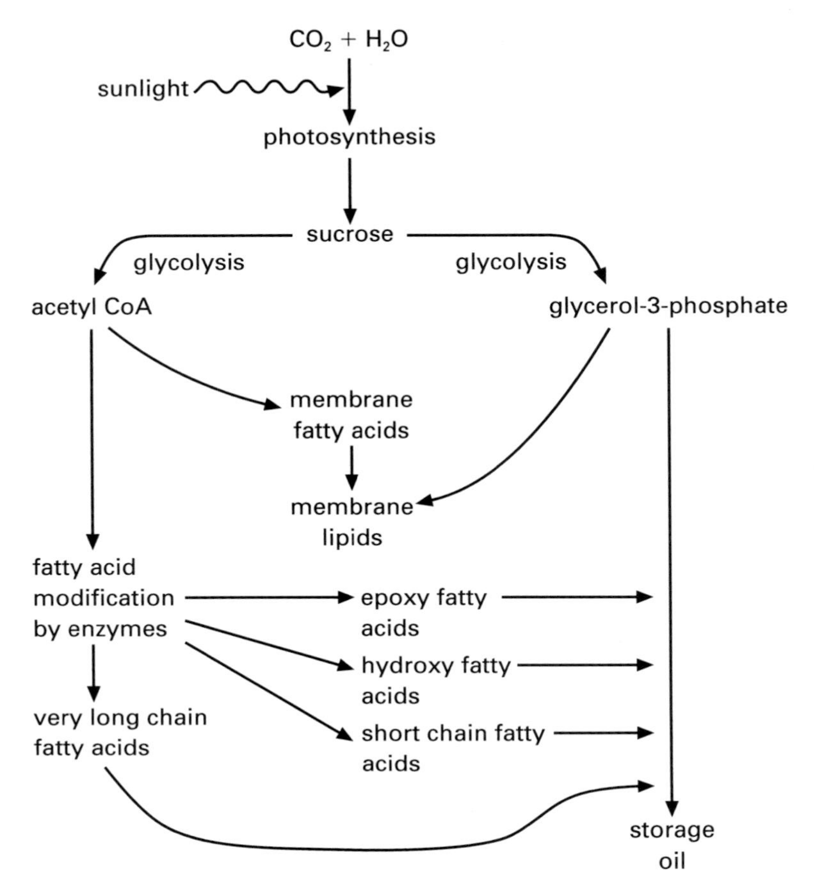

Figure 20.2

(e) Describe the molecular structure of a fatty acid. **[2]**

(f) What type of lipids are found in membranes? How are these lipids suited to their role as a structural component of a membrane? **[4]**

(g) What are the advantages of using vegetable oils as a source of fuel compared with using mineral oil derived from fossil fuels. **[3]**

Some oil-producing crops are able to produce a range of fatty acids because they contain modification enzymes such as hydroxylases and elongases. These enzymes alter the length of the fatty acid chains. However, sunflowers and oil seed rape lack this enzyme. They produce seed oil that is 90% 18-carbon fatty acids which make edible oils. However, it will be possible to modify these crops to produce a wider range of oils for different purposes.

(h) Describe the process by which a sunflower could be modified to produce a different type of oil. **[4]**

Synoptic essays

The synoptic essay is designed to test a candidate's ability to describe and explain biological processes and examine their understanding of biological principles and concepts. Candidates will be expected to develop an argument and present evidence for and against a statement and show that they are aware of the implications and applications in modern biology. Marks will be awarded for the style of presentation, the selection of material and the quality of written communication.

Essay writing – some guidance

An obvious starting point – read the question carefully. It is surprising how many candidates misread the question, for example mixing up the terms 'desertification' and 'deforestation'.

Before you start writing you need to jot down all of your ideas. Organise these ideas so that they form a logical sequence. This forms your essay plan. Synoptic essays are expected to cover a range of ideas from all parts of the specification. So check your plan and make sure it includes material from AS and A2. It is important to get a good balance of information. For example, if the question is about living organisms make sure you cover a wide range of organisms and not just mammals. If the question specifies plants and animal make sure you give equal weighting to both. Don't write an essay on animals and then tack a paragraph about plants on at the end.

Essays need to be structured. They start with an introductory paragraph. This 'sets the scene' and introduces the subject matter. This is followed by the main body of the essay where you present your discussion or argument. The essay is finished with a concluding paragraph. While you are writing, it's a good idea to refer back to your plan to make sure you are still on the right track. It is easy to stray into another subject area when you are writing quickly. Tick off the subjects on your plan so you do not repeat yourself or miss anything out.

When answering these synoptic essays, you will be expected to discuss the subject matter more fully and at greater depth than in shorter free-prose style answers. You will also be judged on your communication skills. This includes your style of writing, which should be well organised, legible, with good grammar and spelling.

Example essay titles

20.3 The role of ATP in plants and animals **[15]**

20.4 The central role of DNA in living organisms **[15]**

20.5 The importance of water to living organisms **[15]**

20.6 The functions of carbohydrates **[15]**

The mark schemes for these essays are not detailed, but are designed to give you an idea of the types of subjects that could be included in the essay.

21 Data handling

Biologists are expected to have a number of mathematical skills. Some of these will be tested in practical sessions, but you can expect to be given mathematical problems to solve in questions, for example working out the magnification of a drawing, calculating the percentage increase or rate of increase and completing statistical tests.

You will be expected to be able to:
- find an arithmetical mean
- know the difference between the terms mean, median and mode
- construct and interpret frequency tables and diagrams
- construct bar charts and histograms
- plot a graph showing two variables using data obtained from an experiment or other source
- calculate the rate of change from a graph showing a linear relationship
- understand probability to work out genetic ratios
- understand the principles of sampling
- understand the importance of chance
- use a scatter diagram to identify a correlation between two variables
- use a simple statistic test (for example *t*-test, chi squared)
- translate information between graphical, numerical and algebraic forms

What's the difference between the mean, median and mode?
The following 10 values were obtained in an investigation: 6, 8, 9, 4, 7, 4, 4, 5, 6, 8

To calculate the **mean**, all the values are added up and divided by 10

$$\text{mean} = 61 \div 10 = 6.1$$

To determine the **median**, arrange the values in either ascending or descending order. The median is the middle value

4, 4, 4, 5, 6, 6, 7, 8, 8, 9

$$\text{median} = \text{the middle number} = 6$$

The **mode** is the value that occurs most frequently. In the sample set of 10 values, the number 4 occurs three times so 4 is the mode.

Frequency tables

Investigations produce data which can be presented in a table. The typical table consists of columns which display the numerical values of the variables. Each column has a heading which indicates the type of measurement together with the SI units. The measurement is separated from the units by a solidus, for example, Temperature / °C and Time / days. When there is more than one column in a table, the first column displays the data for the independent variable and subsequent columns for the dependent variables. Remember the independent variable is the one that has been selected by the person carrying out the investigation, for example, in an enzyme experiment, the investigator may have selected the temperatures at which the investigations are carried out. The dependent variable is the column with the values that were obtained during the investigation. If the data are carefully laid out in a table it is easy to use the data to plot graphs or to make further calculations. The table below shows a set of data obtained in an investigation to determine the time taken for a substrate to disappear at different temperatures. The rate of reaction is the reciprocal of the time taken for the disappearance, that is $\frac{1}{\text{time}}$

Temperature / °C	*Time taken for substrate to disappear / min*	*Rate of reaction / min^{-1}*
10	29	0.03
20	16	0.06
30	8	0.13
40	5	0.20
50	9	0.11

Bar charts and histograms

Bar charts are used when one or more of the variables is not numerical, for example, with different types of insect, types of vegetable, or blood groups. The bars are of equal width and they do not touch. When the bars are vertical, the chart is sometimes called a column chart. Bar charts are used to plot frequency distribution with discrete data, such as the vitamin C content of different fruits as seen in Figure 21.1.

A histogram is a special form of bar chart. It is used to plot frequency distribution with continuous data. It would be used, for example, to show the variability of leaf length (Figure 21.2). The bars are drawn in ascending or descending order and they should touch.

Graphs

Line graphs show the relationship between two variables, such as rate of reaction against time. The dependent variable is plotted on the vertical or y axis and the

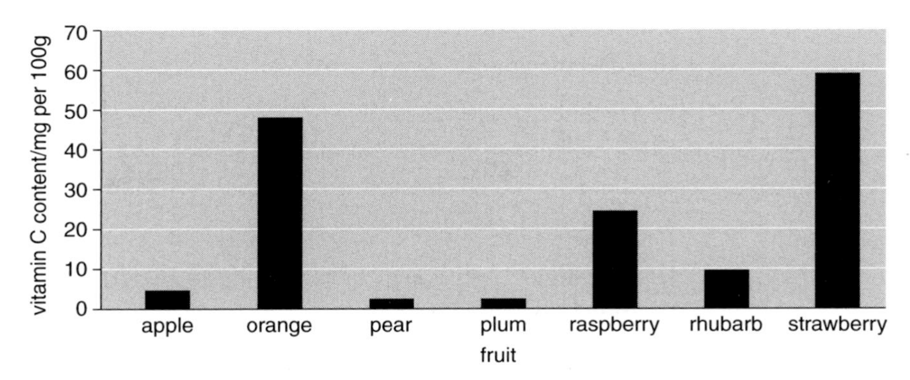

Figure 21.1

independent variable on the horizontal or x axis. Time is normally plotted on the x axis.

All graphs should have a descriptive title. The axes should be drawn the right way round and have labels and units. The scale used to plot the data should be chosen carefully to ensure that the graph fits on the paper and is large enough to be read easily. All the points should be plotted accurately. The points are then joined either by a smooth curve or as series of straight lines joining the points. In most cases, it is better to join the points with straight lines because you do not know how the values vary between the recorded points. The curve is best drawn using a sharp pencil so that the line is clear and errors can be corrected. Use a ruler to

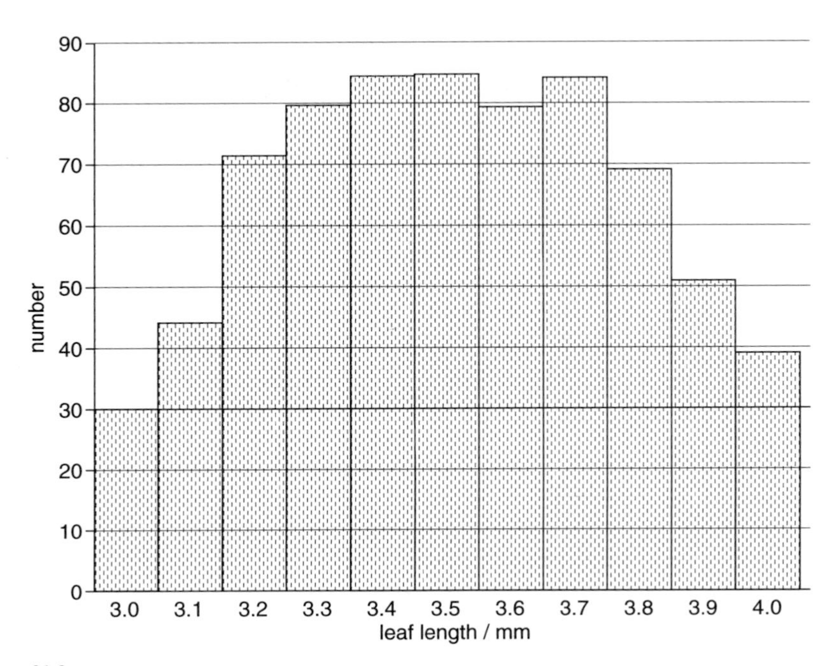

Figure 21.2

produce the straight lines. If there is more than one curve, a key needs to be added to the graph.

You will be expected to be able to plot a graph showing two variables using data obtained from an experiment or other source. In addition, you may be asked to calculate the rate of change from a graph showing a linear relationship. This can be calculated from the gradient of the curve.

For example, the table shows the distance moved by bubbles in a potometer with two different leaf samples

Time / min	*Distance moved by bubble / mm*	
	Leaf A	*Leaf B*
2	15	5
4	60	7
6	110	15
8	140	22
10	170	25
12	182	28
14	200	33
16	220	37
18	240	44
20	275	56

Figure 21.3 shows the data in graphical form.

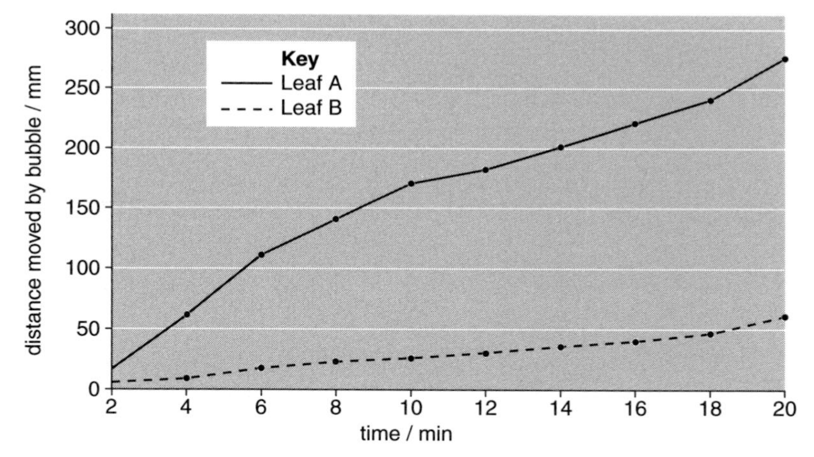

Figure 21.3

Simple statistic tests

There are two simple tests which you may be expected to use in a written examination, chi-squared and the *t*-test. You would not be expected to work out the answer from scratch, but would be provided with all the formulae.

The null hypothesis
In order to use a statistical method to test for an effect, it is necessary to make a hypothesis. This is just an idea and the aim of the investigation is to produce evidence that will prove or disprove the hypothesis. Often it is difficult to actually prove a hypothesis, but it is much easier to find evidence that will disprove it. Therefore, it is usual to make a negative, or null, hypothesis. For example, 'there is no difference in the number of oak trees found in two different woods'. If, once the data are analysed, the null hypothesis is no longer acceptable, then there must be a difference. The null hypothesis has to be rejected.

Chi-squared
The chi-squared, χ^2 is a statistic that is used to test an association between two sets of measurements collected in different places (frequency data). The data must be grouped into classes, for example colour, size, sex and the total number of observations must exceed 20. This test is most frequently used in genetics investigations. It has limited use in ecological investigations.

Chi-squared is a significance test. It tells us whether a relationship is significant or not, and it is based on probability or chance. In order for the relationship to be significant the probability of the results being due to chance must be low. The acceptable level is 5% significance. This means that results due to chance occur only 5 times in every 100. It is also possible to use higher significance levels, such as 1% or 0.1%.

A null hypothesis states that there is no relationship and that differences are due to chance. The data are tabulated in the following manner:

groups	*observed frequency* (O)	*expected frequency* (E)	*difference* (O − E)	$(O - E)^2 / E$

$\chi^2 = \Sigma\,(O - E)^2 / E$ (where Σ = sum of)

The value of E is calculated by dividing the total number in the observed frequency column by the number of groups. The degrees of freedom are calculated by subtracting 1 from the number of groups, i.e. $n - 1$. A significance level is selected and the χ^2 value is read from a table using the degrees of freedom. The null hypothesis can be rejected if the calculated χ^2 is greater than the critical value.

Worked examples
Worked example 1
A student had read that the smooth periwinkle, *Littorina obtusata*, was more likely to be found on bladder wrack than on other seaweeds. In order to test this preference, the student located 100 smooth periwinkles in a 25 m^2 sample area of rocky shore. For each periwinkle, the student noted the seaweed on which it was found.

Null hypothesis: there is no difference in the numbers of smooth periwinkles on the different seaweeds.

groups	*observed frequency* (O)	*expected frequency* (E)	(O − E)	$(O - E)^2 / E$
spiral wrack	2	25	23	21.16
egg wrack	30	25	5	1
bladder wrack	61	25	36	51.84
serrated wrack	7	25	18	12.96

[Note: ignore negative signs]

If the periwinkles were randomly distributed the student would expect to find 25 periwinkles on each of the seaweeds.

$\chi^2 = \Sigma\ (O - E)^2 / E$ (where Σ = sum of)
$= 86.96$

Degrees of freedom $(n - 1) = 3$

Referring to a reference table of chi-squared values, the critical value of chi-squared at the 5% level is 7.82 and at the 1% level is 11.34. The calculated value of 86.96 exceeds this value, so the null hypothesis can be rejected. The periwinkles are not randomly distributed, but showed a preference for certain seaweeds.

Worked example 2

Chi-squared is frequently used to analyse the results of genetics investigations. In an investigation, a pure breeding normal-leafed, purple-stemmed tomato was crossed with a potato-leafed, green-stemmed tomato. All of the F_1 offspring had normal leaves and purple stems. One of these offspring was crossed with a potato-leafed, green-stemmed tomato. The results for the F_2 are given below.

The hypothesis stated that the genes were not linked.

Appearance / phenotype	*Observed number* (O)	*Expected number* (E)	*Difference* (O − E)	$(O - E)^2 / E$
normal leaves and purple stems	81	78.25	2.75	0.096
normal leaves and green stems	75	78.25	3.25	0.135
potato leaves and purple stems	72	78.25	6.25	0.499
potato leaves and green stems	85	78.25	6.75	0.582

The second cross was a test or back cross, so it was expected that the ratio in the F_2 would be 1:1:1:1. The expected numbers are calculated by adding up the observed numbers and dividing by 4.

$\chi^2 = \Sigma\,(O - E)^2 / E$ (where Σ = sum of)
$= 0.745$

The degrees of freedom are $4 - 1 = 3$

The value of χ^2 is looked up in a table in order to obtain the probability level. The appropriate line in the table is given below

χ^2	1.01	2.34	4.64	6.25	7.82	11.34
P(%)	80	50	20	10	5	1

A value of 0.745 is lower than 1.01 with a probability of 80%. Since the chi-squared figure is well below the 5% level, the null hypothesis can be accepted – the genes are not linked. Any differences in the results are due to chance alone.

The *t*-test

A *t*-test is carried out to measure the degree of overlap between two sets of data by comparing the two means. The data must be normally distributed around the means in order to use this test.

The value of *t* is large when there is a large difference between the means of the two samples and the data are clustered tightly around these means. However, the value of *t* will be small if the difference between the two means is small and the data are widely spread. So, the larger the value of *t*, the more certain we are that the two sets of data are different.

The value of *t* is checked against a table of critical values. There are three levels of significance although the more usual is the 5% significance level. This means that there is 1 in 20 chance of obtaining a value of *t* equal to the critical value purely due to random variation. If the value of *t* is greater than the critical value, then there is a significant difference between the two sets of data.

Worked example 3

A student investigated the variation in the length of mussel shells on two locations on a rocky shore. A *t*-test was carried out to see if there was any significant difference between the two samples.

Null hypothesis: there was no difference in the length of the shell.

n *(number)*	*Shell length for group A / mm,* x_1	x_1^2	*Shell length for group B / mm,* x_2	x_2^2
1	46	2116	23	529
2	50	2500	28	784
3	45	2025	41	1681
4	45	1936	31	961
5	63	3969	26	676
6	57	3249	33	1089
7	65	4225	35	1225
8	73	5329	21	441
9	55	3025	38	1444
10	79	6241	30	900
11	62	3844	36	1296
12	59	3481	38	1444
13	71	5041	45	2025
14	68	4624	28	784
15	77	5929	42	1764
total for each column	Σx_1 915	Σx_1^2 57 534	Σx_2 495	Σx_2^2 17 043
calculate mean by dividing by n, where $n = 15$	$\Sigma x_1/n = \bar{x}_1$ 61	$\Sigma x_1^2/n$ 3835.6	$\Sigma x_2/n = \bar{x}_2$ 33	$\Sigma x_2^2/n$ 1136.2
mean squared	$\bar{x}_1^2$ 3721	–	$\bar{x}_2^2$ 1089	–
calculate the variance (s^2) $\Sigma x^2/n - \bar{x}^2$	3835.6 − 3721 $s_1^2 = 114.6$		1136.2 − 1089 $s_2^2 = 47.2$	

$$\text{the value of } t = \frac{(\bar{x}_1 - \bar{x}_2)}{\sqrt{\dfrac{s_1^2}{n_1} + \dfrac{s_2^2}{n_2}}}$$

$$\frac{61 - 33}{\sqrt{\dfrac{114.6}{15} + \dfrac{47.2}{15}}} = 8.53$$

This value is checked against the critical values for the t-test. The degrees of freedom for this investigation are calculated as follows:

$$DF = n_1 + n_2 - 2$$
$$= 15 + 15 - 2 = 28$$

The critical value at $p = 0.05$ is 2.05. Since the value of t is greater than this value, the null hypothesis can be rejected.

Other statistical tests that you might use in practical investigations

Mann–Whitney U test

This test is used when the data are not normally distributed. So the median (the middle value in the range) value is used rather than the mean. The Mann–Whitney U test compares the median of two sets of data.

All the data from both samples are arranged in numerical order and ranked, the lowest value having the lowest rank and the highest value the highest rank. Identical values are given the same rank. The ranks for each set of data are added together and the resulting values inserted into the statistical formula. The smallest U values are obtained when there is no overlap between the two sets of data.

Worked example 4

A student investigated the number of mayfly nymphs in two different habitats, a shallow pool and a deep pool in the same river. He kick sampled for one minute at 6 randomly selected sites in each habitat type and obtained the results below.

Sample	*1*	*2*	*3*	*4*	*5*	*6*	*Median*
Shallow	10	9	15	16	21	9	12.5
Deep	2	3	5	12	6	8	5.5

First arrange the data in numerical order and then in rank order. The lowest value is given the lowest rank, the highest value the highest rank.

Rank						6.5	6.5	8	10	11	12
No. in shallow pool						9	9	10	15	16	21
No. in deep pool	2	3	5	6	8			12			
Rank	1	2	3	4	5			9			

Sum the ranks for each set of data.

$$\Sigma R_1 = 6.5 + 6.5 + 8 + 10 + 11 + 12 = 54$$

$$\Sigma R_2 = 1 + 2 + 3 + 4 + 5 + 9 = 24$$

Calculate U_1 and U_2 values using the following formula

$$U_1 = n_1 \times n_2 + \tfrac{1}{2}n_2\,(n_2 + 1) - \Sigma R_2$$
$$= 36 + 3 \times 7 - 24 = 33$$

$$U_2 = n_1 \times n_2 + \tfrac{1}{2}n_1\,(n_1 + 1) - \Sigma R_1$$
$$= 36 + 3 \times 7 - 54 = 3$$

where n_1 and n_2 are the numbers of samples

Using the smaller of the two values, in this case, U_2 of 3, look up the value in the tables of *U* statistics. The critical value at 5% is 5. Since the smallest *U* value is less than that in the table, the difference between the two samples of mayflies was significant at the 5% level.

Correlation

Correlation is a statistical method that answers the question 'Are these two variables associated?' In other words, if one variable changes, the other changes too. For example, a relationship between the soil pH and the percentage cover of a particular plant species.

The Spearman's rank correlation test determines whether the relationship between two variables is significant. The correlation coefficient, *r*, has to be calculated. It can range from 0 when there is no relationship, to +1 when there is a perfect positive correlation and −1 when there is a perfect negative correlation. A positive correlation indicates that as one variable increases, the other increases too. A negative correlation shows an increase in one variable with a decrease in the other.

By plotting the data on a scattergraph, it is possible to tell whether it is worth testing the correlation. The data shown in Figure 21.4 show a relationship. As the velocity of the water in the stream increases, the number of mayflies also increases. It shows a positive correlation and it would be worth carrying out the Spearman's Rank correlation test. A line of best fit can be calculated using regression analysis. If the data fall into a 'U' shape then there is no correlation.

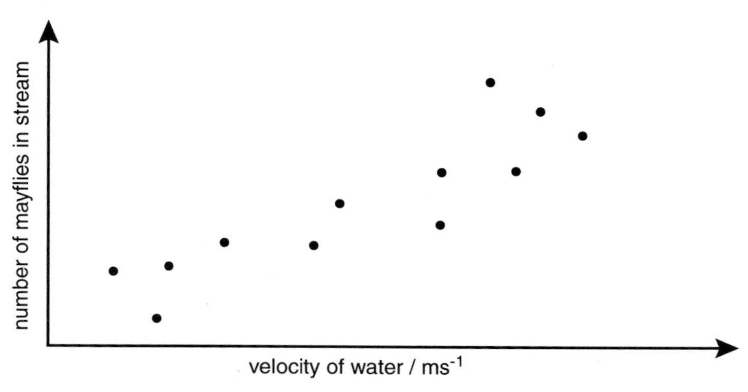

Figure 21.4

Diversity index

The diversity index is a figure which indicates the number of different species present in a particular habitat.

The simplest method of estimating diversity is to collect a sample of approximately 200 specimens and examine each in turn. The specimens do not have to be identified, just given a letter or number. You examine each specimen in turn and you score it as to whether it is the same or a different species from the previous specimen. The index of diversity is calculated by dividing the number of differences by the number of organisms.

Worked example 5

Five different species were found – A, B, C, D, E in the following order

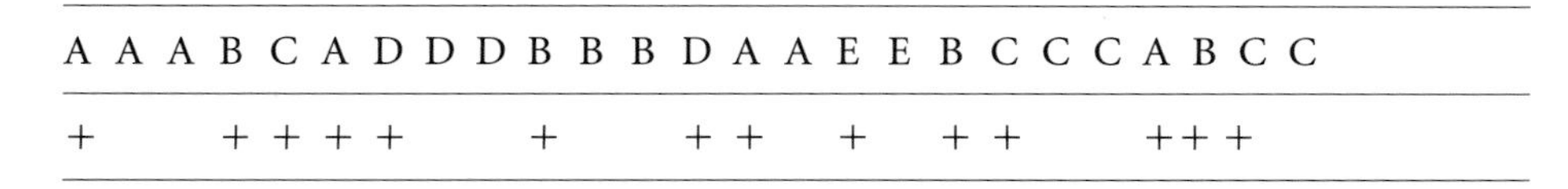

A A A B C A D D D B B B D A A E E B C C C A B C C

+   + + + +   +   + +   +   + +   + + +

The + indicates that the species was different from the previous specimen.

$$\text{diversity index} = 14/25 = 0.56$$

The value obtained in this sample could be compared with that obtained from a different habitat.

Another method is called the Simpson's diversity index. This is commonly used to assess plant diversity. All the plants within the sample area are identified and counted. The diversity is calculated as follows:

$$D = N(N-1)/\Sigma n(n-1)$$

D = Diversity index
N = total number of plants
n = number of individuals per species
Σ = sum of

Sampling

It is not possible to count every single plant of the same species in a habitat. This would take far too much time and cause damage to the habitat. It is, therefore, necessary to work out how much of the habitat needs to be sampled. The study area has to be divided up into small sample areas of equal size. These sample areas have to be selected at random. The data collected from the sample areas can then be used to determine the frequency and distribution of the species across the habitat as a whole.

There are three ways of sampling a chosen area: random, systematic and stratified.

(1) Random sampling. Samples are taken at randomly selected sample sites throughout the area. Since it is difficult to choose the sampling points without any bias, the preferred method is to use random numbers, either from a table or produced by a computer, to generate co-ordinates on a grid. The sides of the grid can be marked out on the ground using two measuring tapes.

(2) Systematic sampling. The samples are taken at regular intervals, ensuring complete coverage of the area. This method is simple to carry out, since it is not necessary to set up a grid with co-ordinates.

(3) Stratified sampling. Often there are differences within the sample area, for example a small area of shrub within a grassland sample or a heavily walked path across the sample area. Therefore, the area should be subdivided and a proportionate number of the samples taken from each of the sub-areas.

Answers

Notes on mark schemes

The answer sections contain mark schemes for the all the questions in this book. Space does not allow for all the possible answers to be listed. Only the most likely alternatives are included. Often there are more suggested marking points than the number of marks available. This means that you can answer the question in a number of different ways and still get full marks. Some answers are followed by notes which draw your attention to difficult points or common mistakes.

Key

; indicates one mark
/ alternative answer
[ ] number of marks

1 Cells

1.1 the basic unit of life; surrounded by a cell membrane; contains organelles; **[2]**

1.2 plant cell larger; plant cell has a cell wall; chloroplasts; large central vacuole; a tonoplast surrounds the vacuole; plant cell has a more regular shape; **[4]**

1.3 **(a)** A – mitochondrion; B – smooth endoplasmic reticulum; C – nuclear membrane; D – nucleoplasm; E – nucleolus; F – rough endoplasmic reticulum; G – Golgi apparatus; H – cell surface membrane / plasma membrane; J – cell wall, K – chloroplast; L – cytoplasm; M – vacuole; N – lysosome; **[13]**

(b) B – synthesis of lipids and steroids; C – controls entry and exit to nucleus; G – synthesis of glycoprotein / modification of proteins; K – site of photosynthesis / production of carbohydrates / production of oxygen; N – production of hydrolytic enzymes / destroying unwanted cell organelles; **[5]**

(c) cellulose; **[1]**

1.4 group of cells carrying out similar functions; lung / liver / blood / bone / connective / epithelial; **[3]**
(Note: there are many other tissue examples)

1.5 tissues are composed of similar cells / organs contain several different tissues; **[1]**

1.6 **(a)** mitochondrion; **[1]**

(b) site of aerobic respiration; **[1]**

(c) width on photograph = 15 mm
(depending on your ruler your measurement could lie between 14 and 16 mm)
15/12 000; = 1.25×10^{-3} mm or 0.00125 mm or 1.25 μm;

Note: when calculating the actual size, divide your measurement by the magnification. Keep to the same units throughout to avoid errors when converting from one unit to another. In general, it is best to use mm. It not really appropriate to give measurements of organelles in metres.

1.7 rough endoplasmic reticulum is encrusted with ribosomes / is concerned with the transport of proteins; smooth endoplasmic reticulum has no ribosomes / is concerned with the synthesis of lipids and steroids; **[2]**

1.8 protein passes into / transported in endoplasmic reticulum; vesicles containing protein are pinched off endoplasmic reticulum; vesicles fuse with Golgi apparatus; proteins modified / carbohydrate added to the them; vesicles containing modified protein pinched off Golgi apparatus; vesicles fuse with cell membrane and release contents; **[4]**

1.9 gives structure to the cytoplasm; keeps organelles in place; involved with movement and transport within the cell; **[2]**

1.10 turn on light source; clean lens; place slide to be examined on stage under low power objective; looking from the side, rack down the objective until it is a few millimetres above the slide; look down the microscope to focus on the slide under low power objective; move slide so that subject of interest is in the centre of the field of view; move high power objective into place; looking from the side, rack down the objective until is just above the objective but not touching; focus on the slide under high power objective; **[4]**

1.11 see greater detail / see smaller organelles; greater magnification; greater resolving power; **[2]**

1.12 magnification is the number of times a point has been enlarged; resolution is the ability to distinguish between two separate points; points that cannot be separated or resolved appear as one point; if the magnification of a microscope exceeds the limit of resolution any further increase in magnification will not give any more detail; **[3]**

1.13

	Prokaryote	*Eukaryotic animal cell*	*Eukaryotic plant cell*
cell wall is made of polysaccharide	✗	✗	✓
DNA enclosed	✗	✓	✓
mitochondria present	✗	✓	✓
cell membrane present	✓	✓	✓
ER present	✗	✓	✓

[5]

1.14 many mitochondria; large amounts of endoplasmic reticulum / large amounts of Golgi apparatus; invaginated cell membrane / microvilli to increase surface area; **[2]**

1.15 **(a)** to prevent any damage to cell contents; **[1]**

(b) to inactivate enzymes / to prevent them damaging the organelles; **[1]**

(c) nucleus is the heaviest organelle; **[1]**

(d) ribosomes; **[1]**

(e) test the fractions for organelle properties / test for the presence of enzymes associated with specific organelles; **[1]**

1.16 presence of microvilli / brush border to increase surface area; presence of large numbers of mitochondria; **[2]**

1.17 **(a)** see diagram below

(b) Marks are given for the accuracy of the drawing, the scale and proportions, relative position and shapes of structures, the neatness of the lines / good single line outlines

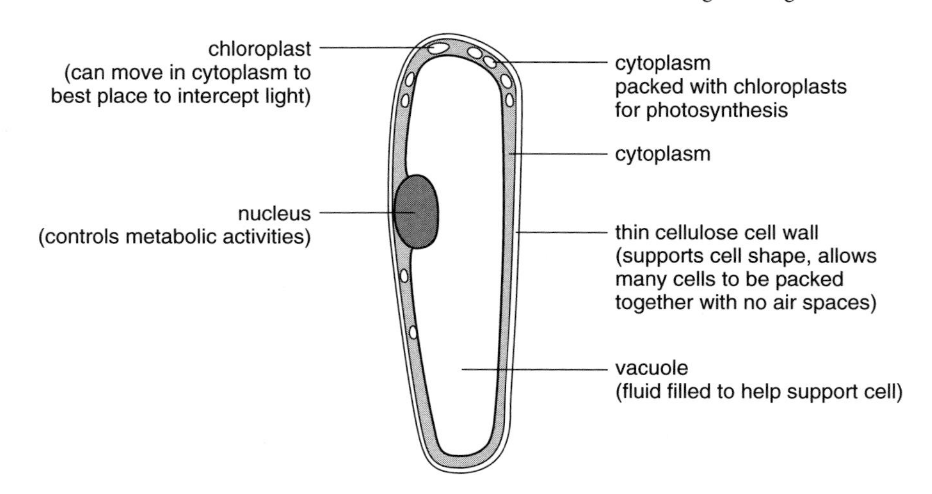

[4]

1.18 using clean cotton wool bud, wipe bud over inside of cheek; smear cells over surface of a clean glass microscope slide; place a couple of drops of stain / methylene blue on the smear; carefully lower cover slip over smear; remove excess liquid with tissue; **[4]**

cell size measured using eyepiece graticule; calibrate this scale with stage micrometer / a microscope slide with a scale marked on it; measure the length of the eye piece graticule against the length of the scale on the micrometer under the high power objective; replace the micrometer with prepared slide; calculate the size of the cells using the graticule scale; **[3]**

2 Membranes

2.1 movement of molecules / ions from a region where they are in higher concentration to one of lower concentration; passive / requires no expenditure of energy; **[2]**

2.2 diffusion of oxygen across alveolar membrane in lungs; diffusion of carbon dioxide into plant leaves; uptake of nutrients from the intestine into the blood system; **[3]**

2.3 movement of water from an area of higher / less negative water potential; to an area of a lower / more negative water potential; through a partially permeable membrane; **[2]**

2.4 permeable to small molecules / water; impermeable to large molecules / sucrose; **[2]**

2.5 uptake of water into root hair cell; **[1]**

2.6 diffusion; active transport; facilitated diffusion; endocytosis; **[4]**

2.7 A – phospholipids; B – proteins; C – glycoprotein; **[3]**
7–8 nm; **[1]**

2.8 hydrophilic – having an attraction for water; hydrophobic – repelled by / not mixing with water; **[2]**

2.9 polar molecules have a charge; **[1]**

2.10 phospholipids have a polar head and non polar tail; non-polar tails face inwards and polar heads point outwards; lipids form a bilayer; **[2]**

2.11 channel proteins to allow facilitated diffusion; carrier protein for active transport of molecules in/out of the cell; receptor molecules for hormones /neurotransmitters; recognition site; enzymes for digestion / respiration; **[2]**

2.12 regulates membrane fluidity; mechanical stability; reduces leakage of polar ions by diffusion; **[2]**

2.13 **(a)** A; (as it has more sucrose molecules) **[1]**
(b) B; (has fewer sucrose molecules) **[1]**
(c) B to A; water moves from a region of higher water potential; to one of lower water potential; **[3]**

2.14 **(a)** the greater the difference in concentration between two regions, the faster the rate of diffusion; **[1]**
(b) small molecules diffuse more quickly than larger molecules; **[1]**

2.15 **(a)** partially permeable membrane; **[1]**
(b)

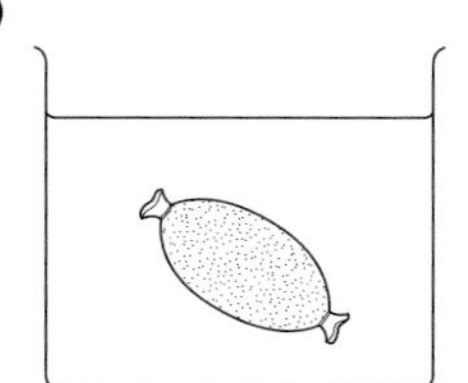

[2]

(c) the sucrose solution inside the tubing has a lower / more negative water potential than the surrounding solution; water moves through partially permeable membrane into tubing; tubing prevents sucrose molecules from moving in the opposite direction; **[2]**

2.16 facilitated diffusion – ions / molecules / sucrose / nucleotides pass through channel proteins in membrane; down concentration gradient; passive / requires no energy / ATP; (example) glucose uptake by red blood cells;

active transport – carrier proteins move ions/molecules across membrane / against the concentration gradient/ from lower to higher concentration; requires energy / ATP; (examples) sodium-potassium pump in cells / calcium pump in muscle cells / reabsorption of amino acids / glucose from filtrate into proximal cell in nephron/ uptake of some molecules from the gut / sodium pump in axon; **[4]**

2.17 **(a)** the surrounding solution more concentrated / has a lower or more negative water potential; water leaves cell; **[2]**
(b) placing cell in a solution of higher water potential / placing in pure water; so water moves back into cell; **[2]**
(c) animal cells do not have a cell wall; when placed in a solution of lower water potential the cell shrinks; **[2]**

2.18 carrier proteins are proteins that span the membrane; transport ions/molecules across the membrane; they are specific for a certain ion/molecule; ions bind to specific receptor site on protein; protein changes shape; releases ion on other side of membrane; **[2]**

2.19 cell surface membrane engulfs/flows around food particle; membrane forms vesicle containing food particle; vesicle pinched off and membrane reforms; **[2]**

3 Biological molecules

3.1 carbon, hydrogen, oxygen; **[1]**

3.2 for energy / used as substrate in cellular respiration; excess carbohydrate converted to glycogen / fat; used in nucleic acids; **[2]**
(Note: these are just a few of the possibilities)

3.3 fish, meat, eggs, milk/cheese, beans/lentils/nuts; **[1]**
(Note: there are more possibilities)

3.4 enzymes; structural (collagen, keratin); carriers in membranes; carry oxygen (haemoglobin); **[3]**

3.5 fat is solid at room temperature, oil is liquid; **[1]**

3.6 a large molecule made up of repeating sub units; which may be identical or similar; joined together to form a long chain; **[2]**

3.7 $C_5H_{12}O_5$; **[1]**

3.8 nitrogen (N); **[1]**

3.9 The completed right-hand column is shown below **[4]**

starch
negative for reducing sugar
non-reducing sugar present, for example, sucrose
soluble protein present

3.10 The missing words are: monosaccharides; dissolve; carbon; condensation; glycosidic; glucose; fructose; glucose; hydrolysis. **[9]**

3.11 **(a)** In (α) glucose, the –OH group on carbon 1 is below the ring; in (β) glucose the –OH group is above the ring;

```
      6CH2OH                          6CH2OH
 H     |       H                  H     |        OH
 |    5C — O   |                  |    5C — O    |
 C  /  OH  H  \ 1C                C  /  OH  H  \ 1C
 |4\   |    |  / |                |4\   |    |  / |
     C3 — 2C                          C3 — 2C
OH    |     |    OH              OH    |     |    H
      H     OH                         H     OH

    α glucose                       β glucose
```

[2]

(b) glycogen is a polymer of (α) glucose; there are 1,4 and 1,6 glycosidic bonds; branched chain; found in liver/muscle/animal cells; cellulose is a polymer of (β) glucose; 1,4 glycosidic bonds only; unbranched chains; found in plants; **[2]**

(c) amylose forms a coil/helix; 1,4 glycosidic bonds; amylopectin has branched structure; 1,4 and 1,6 glycosidic bonds; **[2]**

3.12 amino group has basic properties; carboxyl group has acidic properties; **[2]**

3.13 **(a)**

```
        R1                             R2
H       |      O          H            |       O
  \     |    //             \          |     //
    N — C — C       +         N  —  C — C
  /     |    \             /           |     \
H       |    (OH          H)           |       OH
        H                              H
                    ↓
          R1         H    H
H         |          |    |         O
  \       |          |    |       //
    N  —  C  —  C  —  N  —  C  —  C
  /       |     ||          |       \
H         H     O           R2       OH

                 + H2O
```

[note how the R_2 group rotates]

marks are given for the first line showing the structure of the two amino acids, for the correct linkage of the two amino acids and the elimination of a water molecule. **[3]**

(b) peptide bond; **[1]**

3.14 Coiled polypeptide / α helix; secondary structure; held in place by hydrogen bonds; collagen has several helices / triple helix; some held together by disulphide bridges/bonds; **[3]**
cross links give the structure strength; ability to stretch; withstand pulling force/tensile strength; insoluble; **[2]**

3.15 **(a)** polypeptide chain wound up / folded into a ball shape; tertiary structure; **[2]**

(b) hydrogen bonds; disulphide bridges; ionic bonds; **[2]**

(c) most of the hydrophilic groups are aligned on the outside of molecule; so globular proteins are soluble; makes them more stable; precise shape important for enzymes / antibodies / hormones; **[2]**

(d) iron is part of the haem group; iron bonds with oxygen; **[2]**

(e) breaks hydrogen bonds; affects tertiary structure; loses the precise shape; denaturation; **[2]**

3.16 **(a)** glycerol; fatty acid; **[2]**
(b) ester; **[1]**
(c) one of the fatty acids is replaced with phosphate group; **[1]**
(d) energy; membranes; insulation; metabolic source of water in desert animals; **[2]**

3.17 Benedict's solution turns from blue to a range of colours from yellow to brick red in the presence of reducing sugar;
the depth of colour and volume of precipitate is proportional to the quantity of reducing sugar present in the solution;
take the known glucose solution and make a series of dilutions to produce a range of solutions of known concentration;
carry out a Benedict's test on each, record the colour and volume of precipitate;
make sure each solution stays in Benedict's solution for the same period of time;
carry out tests on each of the unknown solutions and compare to known solutions; **[6]**

3.18 **(a)** $\frac{6}{10}$; = 0.6 (valine); **[2]**
(b) not all present in the sample placed on the paper; some spots too close together to distinguish between them; incomplete hydrolysis of albumen; **[2]**
(c) molecular weight of amino acids varies; each amino acid travels at different speed; reaches different position on chromatogram; **[2]**

3.19 main component of cells / cells up to 95% water;
almost $\frac{3}{4}$ of planet cover by water;
water exists as a liquid at normal temperatures;
provides a medium for molecules and ions to move around;
provided a medium in which life could evolve;
water molecules held together by hydrogen bonds;
oxygen atom has slight negative charge, hydrogen atoms have slight positive charge / dipolar molecule;
creates a weak attraction between the hydrogen and oxygen atoms;
more energy required to break these bonds, so water is a liquid at temperatures where similar molecules are gaseous;
water is a solvent for polar substances;
most cell reactions take place in aqueous solution;
non-polar substances insoluble in water
non-polar substances important in protein and membrane structure;
water is a transport medium in plants and animals;
water has a higher than expected melting and boiling point due to hydrogen bonds;
large quantities of energy required to raise the temperature of a large body of water;
temperature of oceans tends to be constant / stable environmental conditions;
minimises internal changes to body temperature;
large quantity of energy required to evaporate water;
results in effective cooling mechanism in animals;
water needs to lose a lot of energy before it will freeze;
likelihood of freezing reduced / advantage for aquatic organisms;
water as a solid is less dense than liquid water;
ice floats and insulates the water below
allowing animals to survive cold winters;
water molecules are cohesive / tend to stick together;
results in high surface tension;
which is important in transport of water through xylem vessels;
surface tension allows small animals such as pond skater to move across surface of water;
water is involved in hydrolysis reactions and photosynthesis (photolysis); **[8]**

(Note – there are many more points than the number of marks available. In this type of question it is important to cover as many topics as possible, rather than write at length about one topic. In a short account such as this, the marks are more likely to be given for scientific content than style, balance or quality of writing. Try to adopt a writing style that is concise. Do not waste time writing lengthy paragraphs. Think about the point you wish to make and then summarise it in one or two sentences.)

4 Enzymes

4.1 a substance that increases the rate at which reactions take place; does not get involved in the reaction; reused many times; catalysts are used in industrial processes, enzymes are found in living organisms; **[2]**

4.2 polypeptide chain wound / folded; globular /3D shape; held together with hydrogen bonds; ionic; disulphide bridges; **[3]**

4.3 hundreds of different reactions taking place in cell, each enzyme only catalyses one reaction; **[1]**

4.4 part of the enzyme consisting of 3–12 amino acids; with a specific shape; **[2]**

4.5 the active site of the enzyme is the lock; active site has a specific shape; the substrate is the key; only one substrate or group of substrates can fit the active site; **[2]**

4.6 the energy required by molecules to start a reaction; **[1]**

4.7 they hold the substrate/substrates in such a way as to allow them to react more easily; reactions can take place at low temperatures; **[2]**

4.8 heat energy provides molecules with energy; the more heat, the faster the molecules move around; more likely that a substrate molecule will bump into an enzyme; increasing temperatures, increase the rate of reaction; up to a maximum point called the optimum temperature; **[3]**

4.9 high temperatures provide so much energy that the atoms making up the enzyme vibrate; the bonds to break; the enzyme loses its globular shape; denaturation; a few enzymes / bacterial enzymes found in hot springs / industrial enzymes are temperature-resistant; **[3]**

4.10

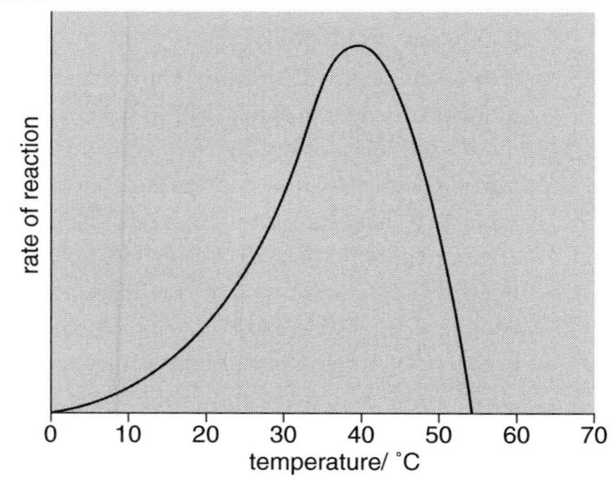

[3]

check that the axes are the right way round, make sure there is an almost straight line from 10–35 °C, a peak and then a rapid decrease in rate, reaching 0 between 50–60 °C

4.11 the rate of reaction doubles for every 10 °C increase in temperature, up to the optimum temperature; **[1]**

4.12 the number of H^+ or OH^- ions in a solution, affects the distribution of charges over the surface of the enzyme / affects ionisation of side chains in amino acid residues; affects the hydrogen bonds; and di-sulphur bridges; which hold the enzyme in its 3D shape; extremes of pH denature the enzyme; **[3]**

4.13 enzymes catalyse different reactions / found in different cellular environments; for example, extra-cellular digestive enzymes have to be able to work in acidic / alkaline conditions in the stomach / intestine; each enzyme has a different combination of amino acid side chains; **[2]**

4.14 a reversible inhibitor binds to the enzyme and reduces its activity; inhibitor can be removed without permanent damage; example, malonate inhibits the enzyme succinate dehydrogenase (involved with Krebs cycle); inhibition reduced by increasing the concentration of the correct substrate; irreversible inhibitor binds to the enzyme permanently; changes the structure of the enzyme and inactivates it; example, arsenic and cyanide permanently damage respiratory enzymes; **[4]**

4.15 non-protein compound needed by the enzyme to function properly; often ions or molecules; may bind to the enzyme so it works more effectively / forms an integral part of the enzyme structure; example, chloride ions and salivary amylase / haem and cytochrome oxidase or catalase; **[4]**

4.16 biological enzymes are denatured at high temperatures; optimum temperature between 30 and 50 °C; **[2]**

4.17 they have optimal temperatures of up to 90 °C; can be used in reactions where high temperatures are involved; they are stable at high temperatures, so can be re-used many times; **[2]**

4.18 **(a)**

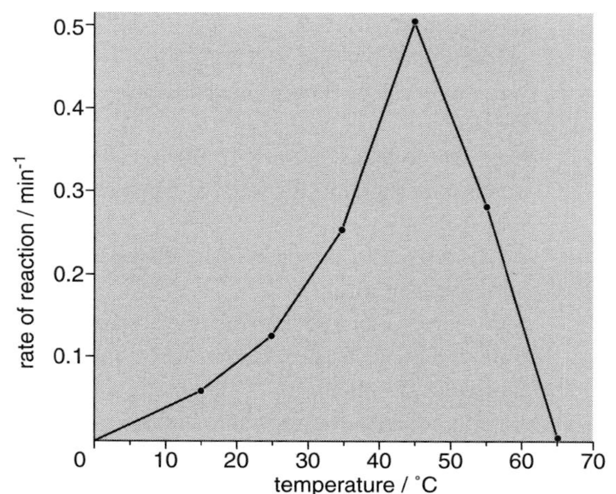

marks are awarded for correctly positioned and labelled axes; suitable scale / size; accurate plotting of points; neat straight line joining up all the points; **[4]**

(b) 45 °C; **[1]**

(c) rate of reaction is increasing with temperature / linear increase; rate doubles with every 10 °C increase in temperature; reference to Q_{10}; **[2]**

(d) rate at 20 °C is 0.1, rate at 30 °C is 0.2; rate of increase is 0.1 /10 °C or 0.01 per °C; **[2]**

(Notes on drawing graphs:
Remember to draw graphs using pencil so errors can be corrected and use a ruler to join the points. Do not draw line of best fit. Lines of best fit can only be applied when it is possible to calculate the relationship mathematically. In graphs displaying biological results, it is only possible to join the points that have been determined in the experiment. For example, in this experiment the rate of reaction was determined at 45, 55 and 65 °C. There is no way of knowing if the rate of reaction peaked between 45 and 55 °C. So, the only option is to join the points with a straight line as an accurate reflection of these experimental results.)

4.19 the rate of reaction is increasing between A and B as there is plenty of substrate; the rate levels off at C because the enzymes' active sites become full; enzyme working at maximum rate; concentration of enzyme is limiting; rate can only be increased if more enzyme is added; **[3]**

4.20 prepare a suspension of yeast cells; make up a range of buffer solutions at pH 4, 5, 6, 7, and 8 / at range of pH values from 4–8; use distilled water as a control; place 5 cm^3 of yeast suspension in test tube; add 10 cm^3 of a buffered pH solution; place in a water bath at 30 °C; leave for several minutes so the temperature in the tube equals that of the surrounding water / equilibrate; add 5 cm^3 of hydrogen peroxide and measure the amount of froth produced over one minute / specified time period; repeat to obtain average; repeat with other tubes; **[4]**

(Note: an acceptable alternative method would be to collect the oxygen via a delivery tube. The volumes in the answer are just suggestions, other volumes could be used.)

Precautions – keep all variables other than the pH constant (i.e. temperature, the concentration and volume of the yeast suspension and the hydrogen peroxide); use same stock of yeast cells; use same method of measurement; **[3]**

4.21 **(a)** proteins; **[1]**

(b) linear increase in activity between pH 4 and 7 / calculation of the percentage increase in activity with each pH unit; reaches optimum / peak at pH 7.5/8.0; activity remains high until pH 9; then activity decreases; decrease in activity not as steep as the increase in activity; **[3]**

(c) increase in activity less steep; enzyme active over greater range of pH values; maximum activity over 4 pH units, which is much larger than that obtained for human digestive enzymes; **[2]**

4.22 **(a)** to prevent the starch altering the pH surrounding the enzyme so prevent denaturation / prevent any reduction in enzyme activity; **[1]**

(b) larger surface area of enzyme exposed to substrate; get higher yields; enzymes can be reused; **[2]**

(c) test with Benedict's solution; if present, blue Bendict's will turn red; **[2]**

(d) use semi-quantitative glucose test strips; changes colour on contact with glucose; colour compared with standardised chart with glucose values; **[2]**

(e) same sized alginate beads used to immobilise enzyme; starch solution runs through at the same speed; test the starch solution for the presence of glucose before starting experiment; column washed with sodium acetate solution before starch added; **[4]**

(f) reduce the size of beads to increase the surface area; increase the speed at which starch flows through column; increase the temperature; **[2]**

5 Heterotrophic nutrition

5.1 A – salivary glands; B – liver; C – stomach; D – pancreas; E – duodenum; F – ileum; G – colon; H – rectum; **[8]**

5.2 hepatic portal vein; **[1]**

5.3 peristalsis; muscles contract and relax squeezing food forwards; swaying movement; **[3]**

5.4 mechanical digestion is the mixing and squeezing of food to churn it up;
chemical digestion involves enzymes; **[2]**

5.5 absorbs water; absorbs mineral ions; **[2]**

5.6 fatty acids; glycerol; **[2]**

5.7

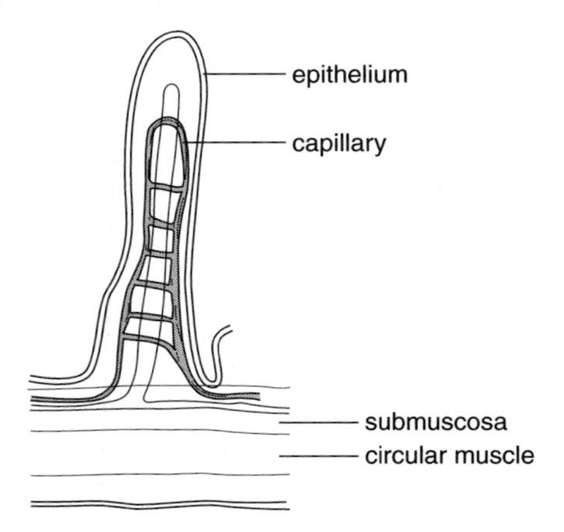

(a) the following structures should have been outlined on your plan – muscle layers; epithelium; capillary network with incoming and outgoing blood vessels; lacteal; **[4]**

(b) four labels correctly placed as on diagram **[2]**

(c) has a large surface area; microvilli to increase surface area further; network of capillaries carries away digested foods; lacteal transports fats; blood supply maintains steep concentration gradient to maximise diffusion; thin epithelium reduces distance for diffusion; smooth muscle allows villus to move in increase absorption; **[4]**

(d) peristalsis / moving food along the gut; **[1]**

5.8

mouth

carbohydrate – containing food placed in mouth; starch digested by salivary amylase, releasing disaccharides;

↓

duodenum

remaining starch digested by pancreatic amylase;

↓

ileum

maltose digested by maltase to form glucose; sucrose digested by sucrase, releasing glucose and fructose;
lacrose digested by lactase, releasing glucose and galactose;

[6]

5.9 the missing words, moving across the rows are: stomach; stomach; caseinogen; trypsin / trypsinogen; duodenum; protein; pancreas; peptides; amino acids;

5.10 acidic pH / 1–2.5 pH; promotes the conversion of pepsinogen to pepsin; **[2]**

5.11 stimulates the secretion of acidic gastric juice; creates conditions for the action of the digestive enzymes / pepsin; **[2]**

5.12 **(a)** hypha/hyphae; **[1]**

(b) extracellular digestion / enzymes secreted onto food; digested food absorbed into hyphae; **[2]**

(c) saprobiontic / saprophytic; **[1]**

(d) saprobiontics feed on dead and decaying organic matter; parasites gain food from a living host; **[2]**

5.13 bile is made in the liver; **[1]**
bile salts; emulsify fats / reduce surface tension of fat droplets; to increase their surface area; so more effectively digested by lipase; **[4]**

5.14 endopeptidase hydrolyses peptide bonds within the protein; produces shorter lengths of polypepetides; exopeptidases break off terminal amino acids; releases single amino acids; **[4]**
endopeptidases released first to break the protein into shorter lengths; producing more terminal amino acids for exopeptidases to work on; **[2]**

5.15 after a meal, concentration gradient exists between ileum and blood capillaries; some amino acids moved by diffusion; diffusion slow / will not move all the amino acids; so rest by active transport; **[2]**

5.16 **(a)** mucosa; **[1]**

(b) secretin causes the release of an alkaline fluid; cholecystokinin causes the release of enzymes; **[2]**

(c) secretin causes the secretion of bile; cholecystokinin stimulates the muscles of the gall bladder to contract; forces bile into the duodenum; **[2]**

(d) inhibits the production of hydrochloric acid in the stomach; delays the emptying of food/chyme into the duodenum; **[1]**

5.17 mutualistic; *Rhizobium* provides plant with supply of ammonium/nitrate ions; plant provides *Rhizobium* with carbohydrates; **[2]**

5.18 rumen contains mutualistic bacteria that secrete cellulase; cellulase digests cellulose; by fermentation; produces carbon dioxide/ methane /organic acids / ethanoic acid; organic acids used by ruminant as source of energy; food may be rechewed or regurgitated; reswallowed for further fermentation; **[3]**

5.19

Feature	*Herbivore*	*Carnivore*
incisors	chisel-shaped / may be absent, for biting grass, upper incisors replaced by horny pad to chew against	small, chisel-shaped to tear food
canines	small, upper ones absent, for nipping off grass, gap behind canines called diastema, a space for the tongue to move the grass to back of mouth	large, pointed, curved to grip prey and tear off flesh
premolars and molars	broad, large grinding surface, ridged, to chew the grass	large, sharp-edged teeth to shear flesh from prey

[6]

6 Exchanges with the environment

6.1 large surface area; permeable; thin; moist; a good transport system; **[4]**

6.2 diffusion over entire surface; oxygen enters; carbon dioxide leaves; along concentration gradients; **[2]**

6.3 diffusion too slow to reach all cells; surface area: volume ratio too small; large animals have higher metabolic rates; **[2]**

6.4 moving air over the gas exchange surface; **[1]**

6.5 A – trachea; B – rib; C – heart; D – diaphragm; E – alveolus; F – bronchus; G – bronchiole; **[7]**

6.6 the missing words are: outwards; external; contracts; increase; decreases; **[5]**

6.7 **(a)** nitrogen gas is not exchanged in the lungs; **[1]**
(b) oxygen in the alveolar air has been taken up into blood capillaries (by diffusion); during expiration, oxygen depleted air mixes with air which has not been in contact with the alveoli; **[2]**
(c) water evaporates from the surfaces of the lungs; **[1]**

6.8 (carbon dioxide in air) enters leaf through stoma; diffuses through the air spaces of the mesophyll; dissolves in surface film of water around palisade cells; diffuses through cell wall and cell surface membrane; **[3]**

6.9 oxygen dissolves in liquid covering alveolar/endothelial cells; oxygen in higher concentration in the alveolus than in blood; oxygen diffuses down concentration gradient; through alveolar/endothelial cell; through capillary wall; **[3]**

6.10 hydrogen ions removed by pump mechanism; potassium ions diffuse into guard cell; pH in guard cell rises; chloride ions diffuse in; malate levels increase; water potential more negative than surrounding cells; water moves into guard cell; turgidity of cell increases; **[4]**

6.11 tidal volume is the volume of air taken into the lungs (during quiet breathing); vital capacity is the maximum volume of air that can be expelled from the lungs by a forceful effort following a maximal inspiration; residual volume is the minimum volume of air that is always in the lungs; **[3]**

6.12 **(a)** 11×0.5 litres $= 5.5$ litres per minute; **[1]**
(b) fills the bronchi/bronchioles/trachea; **[1]**
(c) increases; **[1]**
(d) by 4 times / 400% increase; **[1]**

6.13 $(30 \div 120) \times 100$; $= 25\%$; **[2]**

6.14 **(a)** floor of mouth lowered; pressure in mouth decreases; water drawn in; operculum/gill cover closed; gill cavity increases; floor of mouth raised; water drawn over gills; mouth closes; operculum opens; water forced out through operculum/gillcover; **[3]**
(b) blood flows in opposite direction to water; diffusion gradient between oxygen in water and oxygen in blood always present; more opportunity for oxygen to diffuse into blood; **[2]**

6.15 humans take in air through nose/mouth, insects take in air through spiracles; humans have lungs, insects do not; in humans, oxygen diffuses into blood, in insects oxygen diffuses directly into cells; humans do not have the equivalent to insect tracheoles; humans rely on muscular contractions to draw air into the lungs, most insects rely on diffusion (a few can actively ventilate); **[3]**

6.16 **(a)** inspiratory centre sends impulses to diaphragm and intercostal muscles; diaphragm and intercostal muscles contract; stretch receptors in bronchi/bronchioles send impulses to inspiratory centre; negative feedback control; expiratory centre sends impulses to start expiration; **[3]**
(b) chemoreceptors sensitive to carbon dioxide levels; send impulses to inspiratory centre;

bring about inspiration to lower carbon dioxide levels; **[2]**

(c) increases the rate; and depth of breathing; **[1]**

6.17 **(a)** count the number of ventilation movements of the abdomen in a period of time; change the carbon dioxide concentration by breathing into the tube and count the number of ventilation movements again in the same period of time; **[2]**

(b) use the same locust; repeat the counts at each carbon dioxide concentration; keep all other factors/temperature the same; **[3]**

6.18 **(a)** causes breakdown of the alveoli / reduces surface area available for gas exchange; carcinogens in smoke trigger cancer forming cells; mucus collects in the bronchi leading to inflammation; irritants in smoke cause coughing which can damage the alveoli; **[3]**

(b) in 1916 there were less than 10 deaths per 100 000 from lung cancer, 200 deaths per 100 000 from tuberculosis; steady increase in death from lung cancer from 1925 to 1960, decline in tuberculosis death in same time period; rate of increase of lung cancer greater than rate of decrease of tuberculosis; ×15 or 1500% increase in deaths from lung cancer, tuberculosis fall to less than 10% of the original number; **[3]**

(c) $100 - 35 = 65/35 \times 100$; $= 185.7$ or 186% **[2]**

(d) increased cigarette smoking; more smoking during war years; tuberculosis vaccinations available; better medical treatment of tuberculosis / use of antibiotics; **[2]**

(e) deaths from bronchitis were not recorded before 1935; death rate does not increase due to effective treatment using antibiotics; **[2]**

7 Transport

7.1 large organisms have more cells / have greater requirements for food and oxygen / produce more wastes; too small a surface area : volume ratio to rely on diffusion; need internal transport system to move substances around the body; **[2]**

7.2 the blood passes twice through the heart on each circuit of the body; the pulmonary circulation is kept separate from the body circulation; by a four-chambered heart; **[2]**

7.3 A – aorta; B – hepatic artery; C – hepatic portal vein; D – hepatic vein; E – vena cava; F – renal artery; G – renal vein; H – pulmonary vein; J – pulmonary artery; **[9]**

7.4 **(a)** A – aorta; B – vena cava; C – atrio-ventricular or biscupid valves; D – semi-lunar valves; E – tendinous cords / heart strings; F – pulmonary artery; G – pulmonary vein; **[7]**

(b) left ventricle has to pump blood to the body; right ventricle pumps to lungs; thicker wall pumps blood out with more force; **[2]**

7.5 self exciting / has its own internal or intrinsic rhythm of excitation and contraction; does not require stimulus by nerves; **[2]**

7.6 **(a)** artery diameter measures 25 mm (actual size $= 25/30$; $= 0.8$ mm; (as actual size = size on photo ÷ magnification) **[2]** (make sure you keep to the same units throughout – if you measured in cm the answer will be in cm)

(b) artery has thicker wall than vein; vein has larger lumen; artery has more elastic fibres / smooth muscle; **[3]**

(c) artery – the presence of thicker wall / more smooth muscle / elastic fibres allow artery to withstand the high blood pressure and the surges of blood; elastic fibres allow wall to stretch without damage as blood passes; to recoil afterwards squeezing the blood to create a continuous flow; **[2]**
veins – can hold a larger volume of blood; blood at lower pressure so no need for thick walls; **[2]**

(d) valve; to prevent the back flow of blood; **[2]**

7.7 **(a)** to carry oxygen from the lungs to the tissues; **[1]**

(b) defensive / to engulf foreign bodies and bacteria; **[1]**

(c) red blood cell – no nucleus; creating biconcave shape to allow more haemoglobin to be carried; greater surface area;
neutrophil – can change shape to engulf particles; can squeeze through capillary walls; **[4]**

7.8 tissue fluid – plasma minus red blood cells and plasma proteins; formed when fluid is squeezed out of capillaries under pressure;
plasma – the liquid part of blood; with red blood cells and plasma proteins in suspension; **[3]**

7.9 blood flows into atrium and ventricle; atrio-ventricular valve open; atrium contracts, squeezing remaining blood into ventricle / atrial systole; atrio-ventricular valve closes; ventricle full of blood, ventricular systole / ventricle contracts; blood pushes open semi-lunar valve and passes into artery; semi-lunar valve closes to prevent backflow from artery; atrio-ventricular valve open, atrium and ventricle relax / diastole, heart fills with blood; **[6]**

7.10 sino atrial node or SAN sends out wave of electrical excitation; spreads over atria causing atria to contract /atrial systole; wave reaches AVN; excitation conducted down the bundle of His/through Purkinje tissue; to ventricles; slight delay means that ventricles contract after atria; **[5]**

7.11 **(a)** A – atrio-ventricular valve closes, when pressure in ventricle rises above pressure in the atrium; B – semi-lunar/ aortic valve opens when pressure in the ventricle exceeds pressure in the aorta/artery; C – semi-lunar/ aortic valve closes, when pressure in aorta exceeds pressure in the ventricle; D – atrio-ventricular valves open when venticular pressure falls below that of the atrium; **[4]**

(b) 60/0.8; = 75 beats per minute; **[2]**

7.12 **(a)** haemoglobin carries more oxygen than would be expected due to its high affinity; in conditions of high oxygen partial pressure, the haemoglobin picks up oxygen; oxygen is only released when the oxygen partial pressure levels fall to a relatively low value; **[2]**

(b) **(i)** and **(ii)** compare your curves with the answers shown in the graph **[1 mark each]**

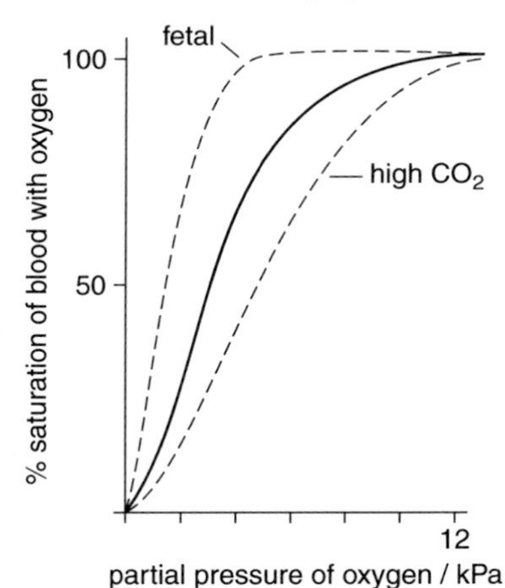

(c) the presence of carbon dioxide causes the curve to shift to the right; reduces the affinity of haemoglobin for oxygen; allows oxygen to be released more easily; fetal haemoglobin has a higher affinity for oxygen than normal haemoglobin; the curve shifts to the left; so the fetal haemoglobin can obtain oxygen from the maternal haemoglobin; **[4]**

7.13 **(a)** there is a constant / linear increase in the number of red blood cells; **[1]**

(b) number of blood cells has increased by approximately 23–26%; **[1]**

(c) red blood cell count continued to increase after maximum altitude had been reached; **[1]**

(d) there is less oxygen in the atmosphere / lower atmospheric pressure; need more red blood cells to carry oxygen; **[2]**

(e) to acclimatise to the altitude; to increase the number of red blood cells; so more red blood cells to carry oxygen during the competition; **[2]**

7.14

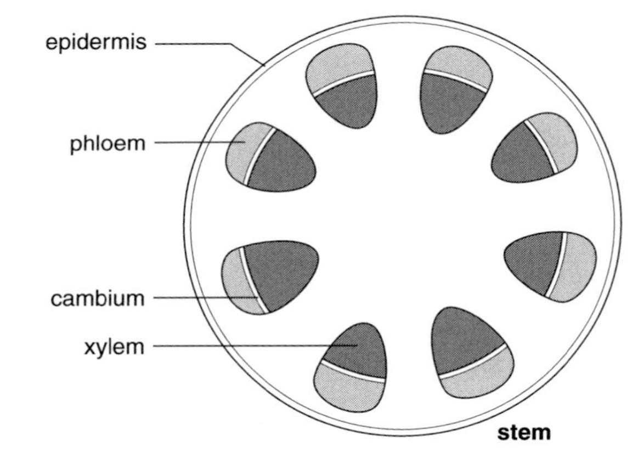

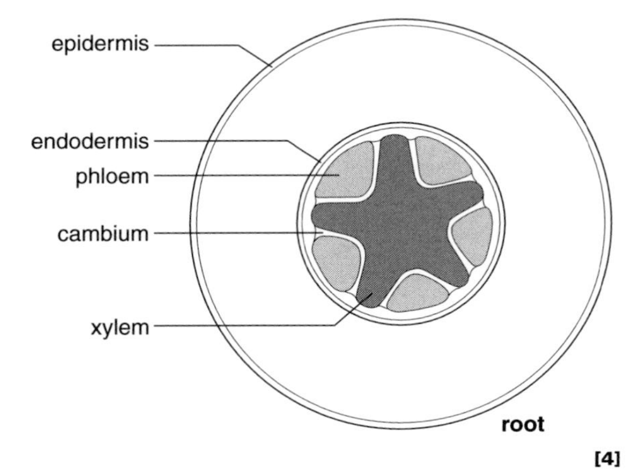

[4]

7.15 osmosis; **[1]**

7.16 diffusion; **[1]**

7.17 dead / no living contents; more room to carry water; lignified walls for support and strength; no end walls to create a continuous tube; **[3]**

7.18 apoplast – water can pass along the permeable cell

walls without moving through the cytoplasm; **[1]**
symplast – water passes into the cytoplasm and moves from cell to cell via plasmodesmata; **[1]**

7.19 endodermal cells have bands of impermeable suberin on cell walls / Casparian strips; prevent water and mineral salts using apoplast pathway to reach xylem tissue; **[1]**

7.20 cohesion is the attraction between water molecules / hydrogen bonds; keeps the water together; adhesion is the attraction of water for the walls of the xylem vessels; these two forces together keep the water in a column as it moves up the xylem vessel; **[4]**

7.21 **(a)** count the number of stomata within a given area and divide the number by the area to give the density:
number of stomata in photo = 7;
leaf area in the photo = (0.16×0.12) 0.02 mm^2
$7 / 0.02 = 350$ stomata per mm^2; **[3]**

(b) to allow gas exchange; carbon dioxide to diffuse into the leaf for photosynthesis; oxygen from photosynthesis to diffuse out; **[2]**

(c) during day / when sun is out; **[1]**

(d) potassium ions diffuse in; hydrogen ions are pumped out at same time; by proton pump; pH rises inside cell / less acidic; this causes influx of chloride ions; malate increases in cell; water potential becomes lower / more negative; water moves into guard cell; **[4]**

7.22 **(a)** carry out experiment under still conditions first; introduce an air bubble into capillary tube by removing end of tube from water; measure distance moved by bubble over set period of time; convert the distance (mm) to volume of water by estimating the radius of the capillary tubing and working out the following equation $\pi r^2 \times$ distance; repeat experiment to obtain a mean; carry out series of experiments at different wind speeds using hair drier or fan set at different speeds or distances from potometer; **[4]**

(b) all conditions the same, except for wind speed / keep temperature / light the same; use the same shoot so leaf area the same; make sure no air bubbles/leaks at the joint between the tubing and the shoot; allow shoot to equilibrate to new conditions before taking readings; **[3]**

7.23 **(a)**

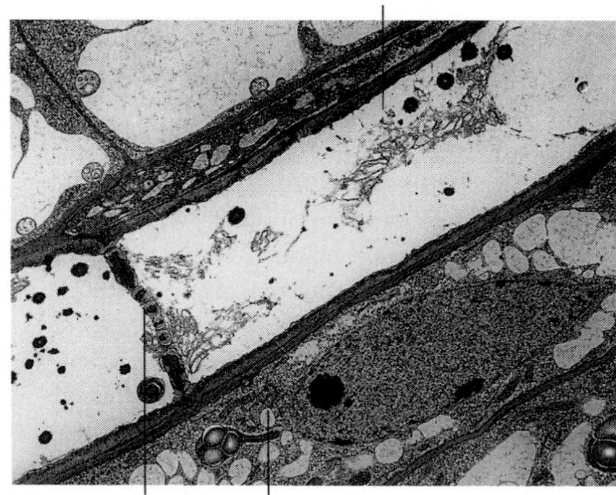

[3]

(b) transport; soluble products of photosynthesis / sucrose / organic molecules; **[2]**

(c) to provide energy / enzymes/ proteins for sieve tube cell; act as transfer cells / move solutes into and out of the sieve tube cell; **[2]**

(d) source is the sieve tubes in the leaves; high hydrostatic pressure; sink is where the sucrose is used / roots / new leaves / seeds; sink has lower hydrostatic pressure; active loading of sucrose into sieve tube; water enters sieve tube cell by osmosis; sucrose unloaded actively at sink; water leaves sieve tubes by osmosis at sink; result is a mass flow of sucrose along sieve tubes; **[3]**

(e) use aphids; allow aphid to insert mouthparts into sieve tube; cut off body and collect exudate; use of radioactive tracers, e.g. $^{14}CO_2$ for use by plant in photosynthesis; leave plant for set period of time; place sections of plant on photographic plate; identify the radioactive tissues; **[3]**

7.24 rolled / coiled leaf / hinge cells; to reduce surface area exposed to environment; to trap humid air to reduce the diffusion gradient / reduce water losses by evaporation;
thick cuticle; to reduce the loss of water by evaporation;
stomata sunken in pits; trap layer of still humid air to reduce evaporation;
hairs; reduce air movement / trap humid air to reduce evaporation; **[6]**

7.25 **(a)** steep / almost linear increase from 06.00–14.00; peak at 14.00; maximum rate is 50 g hr^{-1}; steady fall to 02.00; no transpiration 02.00; **[3]**

(b) evaporation of water from leaf surface; stomata open during the day/closed at night; sunlight raised air temperature; change in wind speed; change in humidity of surrounding air; **[2]**

(c) similar shaped curves; absorption peaks four hours after transpiration; peak of transpiration slightly higher; **[2]**

(d) absorption may not keep up with transpiration / plant would wilt; **[1]**

8 Nucleic acids, genes and gene technology

8.1 DeoxyriboNucleic Acid; **[1]**

8.2 nucleus; **[1]**

8.3 messenger; transfer; ribosomal; **[3]**

8.4 **(a)** A – deoxyribose; B – phosphate; C – hydrogen bonds; **[3]**

(b) adenine; guanine; thymine; cytosine; **[4]**
A links to T; G with C; **[2]**

8.5

DNA	*RNA*
double stranded	single stranded;
sugar is deoxyribose	sugar is ribose;
contains the base thymine	thymine is replaced by uracil;
one type of DNA	three types of RNA;

[4]

8.6 DNA unwinds/untwists; hydrogen bonds between the two strands are broken; bases exposed; pool of free nucleotides in nucleoplasm; complementary bases joined together; by DNA polymerase; new strand forms opposite each original strand; form two new DNA molecules; **[4]**

8.7 **(a)** there are strands that contain both types of nitrogen; as well as strands that contain just the lighter ^{14}N; **[2]**

(b) there would be no mixed strands; there would be a parent strand containing all ^{15}N and daughter strands containing ^{14}N; **[2]**

8.8 genetic code is the order of bases along the DNA; that determines the sequence of amino acids in the polypeptide chain; each triplet codes for a different amino acid; code is based on four bases; there are 64 different combinations of the four bases; more than enough to code for the 20 amino acids; changes in the code cause mutations; **[5]**

8.9 codon is the triplet found in the DNA / messenger RNA molecule; anticodons are complementary to codons; found on the transfer RNA molecule; **[2]**

8.10 **(a)** GGU AUU GUU GAA CAA; **[1]**

(b) CCA TAA CAA CTT GTT;; **[2]**

(c) RNA polymerase; **[1]**

(d) single strand; shape of clover leaf; anti codon; chain of 80 nucleotides; amino acid binding site; 5’-end always ends in CCA; **[2]**

8.11 non-coding region of a gene/DNA; **[1]**

8.12 **(a)** DNA; **[1]**

(b) restriction endonuclease; cuts open the plasmid; at specific nucleotide sequences; **[2]**

(c) TGG and CGT; **[1]**

(d) DNA ligase; joins the ends of the DNA together; **[2]**

(e) recombinant; **[1]**

(f) treat bacterial cells with calcium ions; membranes become permeable; plasmid taken into cell; **[2]**

(g) marker gene; antibiotic resistance; apply antibiotic, those with gene survive; alternative method use radioactive DNA probe; contains part of the desired gene sequence; probe finds cells with the gene; **[2]**

(h) purer; cheaper; larger volumes; engineered insulin produces fewer side effects in patients; **[3]**

8.13 **(a)** 16; **[1]**

(b) DNA polymerase; **[1]**

(c) PCR makes multiple copies of DNA; transcription makes RNA from DNA; **[2]**

(d) can make more copies of tiny samples of DNA; taken from crime scenes / traces of blood / semen / hair follicles; can carry out DNA fingerprinting on samples; **[3]**

8.14 transcription takes place in nucleus; length of DNA unwinds; DNA unzips, using RNA polymerase; one strand of DNA acts as template; molecule of mRNA produced; complementary bases to DNA; mRNA leaves nucleus via nuclear pores; attaches to ribosome in cytoplasm; translation takes place; two codons of mRNA held in position on ribosome; pool of tRNA in cytoplasm; each tRNA molecule holds a different amino acid, as identified by anticodon; anticodon of tRNA pairs with codon on mRNA; peptide bond forms between adjacent amino acids; polypeptide chain forms; polyribosomes allow multiple copies of the same polypeptide chain to form; **[10]**

9 Cell cycle

9.1 prophase, metaphase, anaphase, telophase; **[1]**

9.2 **(a)** A – chromatid; B – centriole; C – spindle; **[3]**
(b) spindle fibres pull chromosomes towards the poles of the cell; **[1]**
(c) anaphase; **[1]**

9.3 haploid – having half the chromosome number / having one of each pair of homologous chromosomes; diploid – having the full number of chromosomes / having twice the haploid number of chromosomes / having two of every chromosome; **[2]**

9.4 **(a)** 46; **[1]**
(b) 23; **[1]**

9.5

Feature	*Mitosis*	*Meiosis*
chromosomes replicate	✓	✓;
involves two nuclear divisions	✗	✓;
occurs in a haploid cell	✓	✗;
there is crossing over between the chromatids	✗	✓;
homologous chromosomes pair up	✗	✓;
sister chromatids are separated	✓	✓;
occurs during growth	✓	✗;
daughter cells are identical to the parent cell	✓	✗;
occurs during gamete formation	✗	✓;

9.6 two chromosomes, one from each parent; same length; centromere is in same position; same number of genes; genes in same order; **[2]**

9.7 growth; repair / replacement of cells/tissues; asexual reproduction in procaryotes / protoctists / fungi; gamete formation in gametophyte generation of plants; **[2]**

9.8 only one parent required; identical offspring; maintains genetic stability; processes such as tillering in grasses and runners in strawberries allow plant to clump up and out-compete other plants; **[2]**

9.9 separation of homologous chromosomes; crossing over in prophase I; independent assortment of chromosomes on spindle at metaphase I; **[3]**

9.10 **(a)** uncontrolled mitosis / division of cells; caused by failure to control the mechanisms of cell division; due to changes/mutations in genes; results in an irregular mass of cells called a tumour; **[2]**
(b) ionising radiation; X-rays; gamma rays; UV light; tar from tobacco smoke; aniline dyes; viral infection / papilloma virus; **[3]**

9.11 **(a)** S; **[1]**
(b) G2; **[1]**
(c) G1 and G2; **[1]**

9.12 in prophase, chromosomes appear; become thicker and shorter; two chromatids visible, held together at centromere; nucleolus disappears; nuclear membrane breaks up; spindle appears; during metaphase, chromosomes become attached to spindle fibres by centromeres; lined up along equator of spindle; at right angles to the poles; during anaphase, spindle fibres become shorter; centromeres divide and are pulled towards the poles; this separates the chromatids; during telophase, chromatids uncoil; are now chromosomes; they are no longer visible; nuclear membrane forms around each group of chromosomes; nucleoli reappear; cytokinesis / division of cytoplasm follows; two cells separated; **[8]**

10 Reproduction

10.1 reproduction involving one parent / no fusing of gametes / reproduction without any genetic variation; examples binary fission in bacteria / protozoa; budding in yeast / hydra; **[3]**

10.2 egg is larger than sperm; large food store / yolk; sperm have head, middle piece and tail; many mitochondria to produce energy for movement; sperm have no food store; sperm have acrosome to breakdown jelly layer around egg;
both have a haploid number of chromosomes; **[3]**

10.3 fewer resources needed to produce a spermatozoan than an egg; to ensure that one sperm makes way to the egg (particularly important in external fertilisation); **[1]**

10.4 identical copy of another organism; by asexual reproduction / vegetative propagation in plants / tissue culture / by removing genetic material from one cell and placing it in an empty egg cell; **[2]**

10.5 **(a)** A – sepals, protect bud; B – anther, produces pollen; C – stigma, receptive to pollen for pollination; D – ovule, contains the female egg / becomes the seed; E – petal, colourful to attract insects; F – ovary, becomes the fruit; **[6]**

(b) large petals; anthers inside flower; large stigma; flower provides platform for insects to land; nectaries to release nectar; **[2]**

(c) small or no petals; no nectaries; stamens dangle outside flower to catch the wind; large, feathery stigma to trap pollen; **[2]**

10.6 small sample of cells removed from donor plant; sterilised; placed on growth medium so that cells increase in number; callus formation; addition of plant growth substances to bring about differentiation of cells into roots and shoots; many plants difficult to reproduce, either by seeds or by vegetative propagation; creating more copies of individuals that may have been produced by genetic engineering (sexual reproduction may result in the loss of the desired gene); **[5]**

10.7

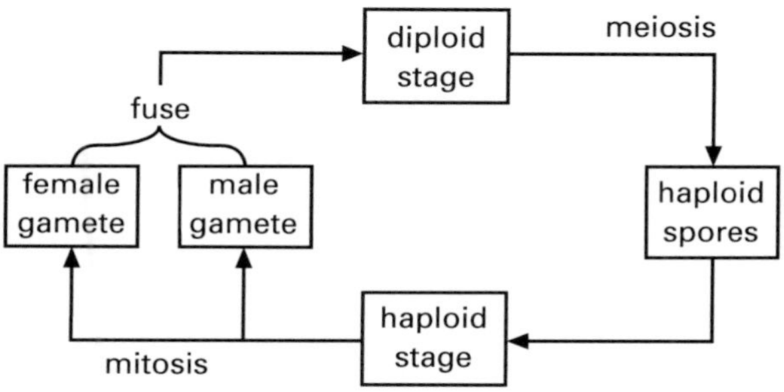

diploid stage
meiosis
haploid spores
haploid stage
mitosis
female gamete
male gamete
fuse

[5]

10.8 protogyny is when the female reproductive parts of the flower ripen first; protandry is when the male parts ripen first; **[2]**

10.9 generative nucleus divides into two; by mitosis; forms two male nuclei; enter embryo sac; one nucleus fuses with egg cell; to form diploid zygote; second nucleus fuses with diploid nucleus to form triploid nucleus; this develops into endosperm; **[3]**

10.10 there are two fusions; one male nucleus fuses with an egg nucleus to form zygote; other fuses with diploid nucleus to form endosperm; **[2]**

10.11 seed absorbs water; testa splits; water activates gibberellin production; which causes release of amylase; amylase hydrolyses starch stores; aerobic respiration; embryo undergoes mitosis; radicle appears; followed by shoot; **[3]**

10.12 A – epididymis; B – seminal vesicle; C – prostate gland; D – vas deferens / sperm duct; E – urethra; F – testis; G – penis; **[7]**
sperm production best at 35 °C; so held outside body where it is cooler; (remember body temperature is 37 °C) **[2]**

10.13

	FSH	*LH*	*Oestrogen*	*Progesterone*
secreted from the pituitary	✓	✓		
repairs the endometrium			✓	
inhibits the release of LH by negative feedback				✓
brings about formation of the corpus luteum		✓		
secreted by the corpus luteum			✓	✓
stimulates the development of a follicle	✓			
brings about ovulation	✓	✓		
maintains the endometrium				✓

[8]

10.14 **(a)** provides large surface area for exchange; maternal blood flows close to the fetal blood; provides a minimal barrier for the diffusion and active transport of materials; **[2]**

(b) blood pressure is higher; this would damage fetal blood vessels; **[2]**

(c) entering water / oxygen / glucose / amino acids / lipids / mineral salts / vitamins / antibodies / hormones; leaving urea / metabolic waste products / carbon dioxide; **[2]**

(d) forms a cushion to protect fetus from knocks / bumps; keeps temperature constant; allows fetus to move; **[1]**

10.15 A – spermatogonium, 2n; B – primary spermatocyte, 2n; C – secondary spermatocyte, n; D – spermatid, n; E – spermatozoan, n; **[5]**

11 Genetics

11.1 allele refers to the alternative forms of a gene; alleles are usually either dominant or recessive; **[2]**
a dominant allele will be expressed in the phenotype; even if there is only one allele present; it will mask the recessive allele; **[2]**
recessive alleles are only expressed if both alleles present are recessive; a single recessive allele will

be masked by a dominant allele; **[2]**
homozygous means that both alleles present in an individual are the same/identical; two dominant or two recessive alleles; **[2]**
heterozygous means that the alleles present in an individual will be different , for example, if R represents dominant and r is recessive, a heterozygous individual will have Rr; **[2]**
codominant means that alleles of a gene are equally dominant to each other; both alleles expressed in phenotype; **[2]**

11.2 **(a)** purple stems; **[1]**

(b) key: P represents purple-stemmed and p represents green-stemmed.

parents	PP × pp
gametes	P × p;
F_1	Pp all purple-stemmed;
F_1 parents	Pp × Pp
gametes	P or p × P or p;
F_2	PP, Pp, Pp, pp;
phenotypic ratio	3 purple : 1 green; **[5]**

(c)

backcross parents	Pp × pp;
gametes	P or p × p;
F_2	Pp and pp;
phenotypic ratio	1 purple : 1 green; **[4]**

(Note: Sometimes genetics questions tell you which letters to use. If they do not, make sure you start off with a key to the meaning of the letters used in your answer. Always write down all the stages, including parents, gametes and resulting offspring. If the question refers to genotype, it requires you to work out the genetic composition of the offspring. If the question asks for phenotype you must indicate the appearance of the individual based on their genotype. Often you are asked to work out the ratio of phenotypes. This is a figure such as 3 : 1. Always make it clear to which phenotype the ratios refer, for example 3 purple-stemmed : 1 green-stemmed.)

11.3 **(a)** key: T is tall, t is short, W is white and w is red.
parental genotype TTWW; ttww; **[2]**

(b) TtWw; tall and white; **[2]**

(c) parents TtWw × ttww;
gametes TW, Tw, tW, tw × tw;

gametes	*TW*	*Tw*	*tW*	*tw*
tw	TtWw tall/white	Ttww tall/red	ttWw short/white	ttww; short/red;

ratio 1 tall/white : 1 tall/red : 1 short/white : 1 short/red; **[5]**

11.4 **(a)** DdHh; dark and short-haired; **[2]**

(b)

gametes	*DH*	*Dh*	*dH*	*dh*
DH	DDHH dark/short	DDHh dark/short	DdHH dark/short	DdHh dark/short
Dh	DDHh dark/short	DDhh dark/long	DdHh dark/short	Ddhh dark/long
dH	DdHH dark/short	DdHh dark/short	ddHH albino/short	ddHh albino/short
dh	DdHh dark/short	Ddhh dark/long	ddHh albino/short	ddhh albino/long

phenotypic ratio 9 dark/short : 3 dark/long : 3 albino/short : 1 albino/long **[6]**

(Mark allocation – 1 mark for correct gametes, 3 marks for correct genotypes (deduct one mark for each error), 1 mark for correct ratio and 1 mark for the correct phenotypes.)

(Note: it is important when filling in a Punnett square with the full 16 combinations that you write down the parental gametes in the correct order. Start with the two dominant alleles, then one dominant and one recessive, then recessive and dominant and finally two recessive alleles, like this AB, Ab, aB, ab. If you get the order correct you can check your workings out very easily. Look on the Punnett Square above. The bottom right hand corner always has the one double recessive genotype. It is surrounded by the three dominant/recessive combinations, in this case albino/short. Look for the position of the three dark/long genotypes. All the rest are the 9 dominant genotypes.)

11.5 **(a)** hairy stems; there are three times as many hairy as smooth stems / 63 : 21 hairy : smooth; **[2]**

(b) parental genotypes HhYy; Hhyy; **[2]**

(c) gametes HY, Hy, hY, hy × Hy, hy;

gametes	*HY*	*Hy*	*hY*	*hy*
Hy	HHYy hairy/ yellow	HHyy hairy/ white	HhYy hairy/ yellow	Hhyy hairy/ white;
hy	HhYy hairy/ yellow	Hhyy hairy/ white	hhYy smooth/ yellow	hhyy smooth/ white;

(d) 6 : 2 or 3 : 1 hairy : smooth; **[1]**
(e) 4 : 4 or 1 : 1 yellow : white; **[1]**

11.6 **(a)** 10 purple/hairy : 3.2 purple/hairless : 1 green/hairy : 1 green/hairless;; **[2]**
(b) 9 : 3 : 3 : 1; as it is a dihybrid cross/ two genes are involved; **[2]**
(c) too few plants; some of the seeds of the recessive genotypes did not germinate; some plants died before their characteristics became apparent; incorrect interpretation of results; some plants did not express their characteristics; **[3]**
(d) carry out statistical test; chi squared test; **[2]**

11.7 No; if the man was blood group AB he could father the children with A and B blood groups but not the group O child; if the father was $I^A I^O$ he could father two of the children, but not the blood group B child; if the father was $I^B I^O$ he could father blood groups O and B, but not blood group A child;

Using the first case:
parent $I^O I^O \times I^A I^B$
gametes $I^O \times I^A$ or I^B;

F_1 $I^O I^A$ or $I^O I^B$;

children could have blood group A or B **[6]**

11.8 **(a)** key: R is red flowers, r is white
parents RR × rr;
gametes R × r;
F_1 Rr all pink; **[3]**
(b) F_1 parents Rr × Rr;
gametes R or r × R or r;
F_2 RR, Rr, Rr, rr;
phenotypes 1 red : 2 pink : 1 white; **[4]**

11.9 A – $X^C Y$; B – $X^C X^c$; C – $X^C X^c$; D – $X^c X^c$; E – $X^c Y$; **[5]**

11.10 key: B is black and b is ginger
parents $X^B X^B \times X^b Y$;
gametes $X^B \times X^b$ or Y;
F_1 cats $X^B X^b$, $X^B Y$;
phenotypes tortoiseshell female and black male; **[4]**

11.11 **(a)** parental genotype PPrr; × ppRR; **[2]**
(b) F_1 PpRr; all walnut; **[2]**
(c) parental genotypes PpRr × PpRr

gametes	*PR*	*Pr*	*pR*	*pr;;*
PR	PPRR walnut	PPRr walnut	PpRR walnut	PpRr walnut
Pr	PPRr walnut	PPrr pea	PpRr walnut	Pprr pea
pR	PpRR walnut	PpRr walnut	ppRR rose	ppRr rose
pr	PpRr walnut	Pprr pea	ppRr rose	pprr single;;

phenotypic ratio 9 walnut : 3 pea : 3 rose; 1 single; **[6]**

11.12 **(a)** genes found on the same chromosome; inherited together; may be separated by crossing over/ chiasmata; **[3]**
(b) 3:1; because the two genes would be inherited as one / inheritance the same as for a monohybrid cross;
9:3:3:1; because the genes would be inherited separately / would be a normal dihybrid cross; **[4]**

12 Classification

12.1 group of individuals with several features in common; can interbreed; produce fertile offspring; **[2]**

12.2 all organisms are given two Latin names; one representing genus; the other species; **[2]**

12.3 to ensure everybody refers to the same organism; often species have a local name that is not recognised by other people; Latin was in common usage / the scientific language when biologists first started to classify organisms; **[2]**

12.4

Kingdom	*Feature*	*Example*
Prokaryotae;	lack membrane-bound nucleus / lack membrane-bound organelles / lack 9 + 2 microtubules;	bacteria such as *Escherichia coli* / photosynthetic cyanobacteria / Anabaena;
Protoctista;	eukaryotic cell / bound nucleus / often unicellular or groups of similar cells;	a named green or brown alga / named, protozoa such as *Amoeba* / *Paramecium*;

Kingdom	*Feature*	*Example*
Fungi;	non-photosynthetic eukaryote / non-cellulose cell wall / mostly multinucleate hyphae / spores without flagella;	named fungus / mucor / *Rhizopus*, *Agaricus*;
Plantae;	multicellular / photosynthetic / cellulose cell walls;	moss / fern / angiosperm / named example of a plant;
Animalia;	multicellular / non-photosynthetic / nervous co-ordination;	named animal such as snail / earthworm / spider / human / dog;

[15]

12.5

Kingdom	Animalia;
Phylum	Chordata;
Class	Mammalia
Order	Primates
Family	Hominidae
Genus;	*Homo*
Species;	*Homo sapiens*

[4]

12.6 refers to a series of groups; starts with the largest group called kingdom; organisms divided into smaller groups / taxons according to their shared characteristics; **[2]**

12.7 phylogeny – evolutionary history; phylogenetic classification based on evolutionary history / relationships between organisms; looks at homologous structures, such as pentadactyl limb; artificial classification based on similarity of features; such as number of legs /number of petals; **[4]**

12.8 examine the visible external features / use a key to group the unknown with other known species; examine the pollen under a microscope and compare to known species; take DNA sample and produce a DNA 'fingerprint' and compare to known species; **[3]**

13 Selection and evolution

13.1 differences between individuals; in a population; examples – height / colour of hair / number of eggs in a clutch of eggs; **[2]**

13.2 crossing over / chiasmata; independent assortment of chromosomes; **[2]**

13.3 eye colour – gene; blood group – gene; hair colour – gene; intelligence – environmental; height – combination of both; weight – both; **[5]**

13.4 continuous – normal distribution; most individuals around the middle values / mean; examples – height / weight; discontinuous – does not display normal distribution; two modes; two distinct groups of individuals; eye colour / blood groups;

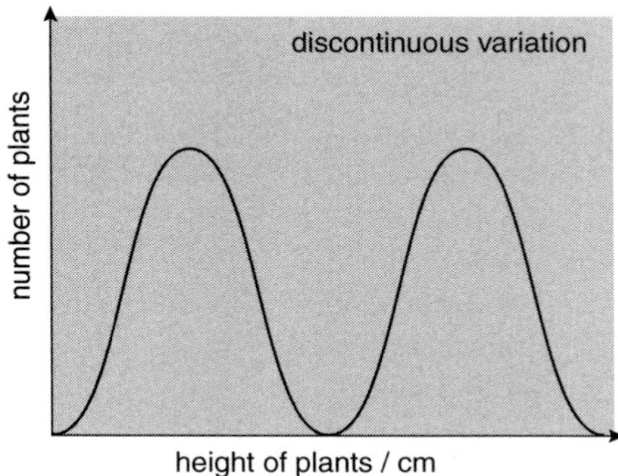

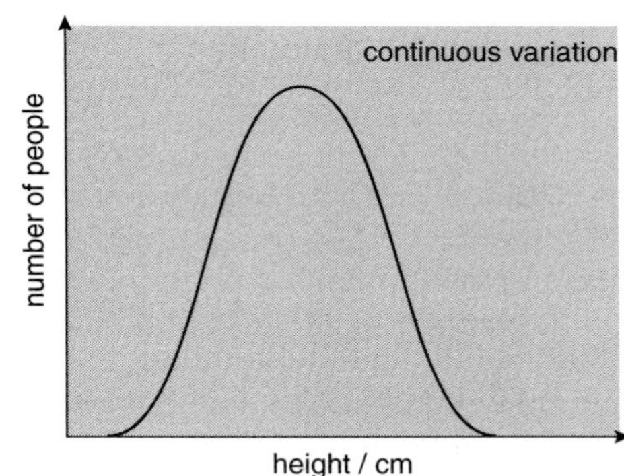

[6]

13.5 formation of a new species; two populations become isolated; cannot reproduce; **[2]**

13.6 barrier such as mountain / river / glacier; separates a population of individuals; each group experiences different selection pressures; divergence between the groups; **[3]**

13.7 **(a)** speciation which occurs within a population; in the same area; reproductive isolation; examples – differences in courtship patterns / differences in reproductive organs / hybrid sterility / polyploidy in plants /species present at different times of year (temporal); **[4]**
sympatric speciation occurs in same area, allopatric speciation occurs in different geographic areas; **[2]**

13.8 **(a)** codon will code for a different amino acid; may or may not have an effect on the final polypeptide structure; **[2]**

(b) frame shift / bases would shift by one position / all subsequent codons would be changed; amino acid sequence would be changed; change in protein structure; **[2]**

13.9 homologous pair of chromosomes fail to separate; both move to the same pole during anaphase; one gamete receives an extra chromosome while the other receives none; can affect every pair of chromosomes / can get complete non-disjunction; in Downs syndrome, the gamete has an extra chromosome 22; the individual has 47 chromosomes; **[4]**

13.10 **(a)** 63; gamete of horse has 32, gamete of donkey has 31; **[2]**

(b) there is an odd number of chromosomes; homologous chromosomes would not pair during meiosis; **[2]**

13.11 **(a)** the total number of genes; in a population; **[2]**

(b) the number of individuals in a population; that carry a particular allele; **[2]**

(c) the exchange of alleles / interbreeding; with members of other populations; **[2]**

13.12 prezygotic – prevention of fertilisation and zygote formation; caused by incompatible genitalia / courtship failure / individuals present at different times of year / seasonality; **[2]**
postzygotic – fertilisation takes place but zygote fails to develop, zygote aborts; / any resulting offspring sterile / hybrid sterile; **[2]**

13.13 height – genetic; as the values of identical twins raised together and apart are the same, but different to the non-identical twins;
mass – environmental; since the values for the identical twins raised together and apart are different;
IQ – environmental; since the values for the identical twins raised together and apart are different; **[6]**

13.14 **(a)** shows continuous variation; majority of population or 75% have a birth weight of 2.5–3.5 kg; 20% of population have the mean birth weight of approx. 3 kg; less than 1% of population have either the lowest or highest birth weights; **[3]**

(b) inverse relationship; at the extremes of birth weight mortality is high; at the mean or most frequent birth weight, mortality is at its lowest; **[2]**

(c) **(i)** poorly developed lungs / needs more food;
(ii) needed more of the mother's resources; problems giving birth; **[2]**

(d) stabilising; **[1]**

13.15 **(a)** period of advantage for the dark forms was short compared with that of the light forms / dark forms only at an advantage at night; longer period of predation; **[2]**

(b) air pollution; darkens the trees; **[2]**

(c) having more than one appearance; still belonging to the same species; **[2]**
peppered moth / banded snail / tropical swallowtail butterflies; **[1]**

13.16 **(a)** normal; **[1]**

(b) Graph a shows normal distribution with a mean activity score of approximately 35; graph has a wide base; graph b shows two distinct curves, one with a mean of less than 5 and the other with a mean score of 55–70; narrower range of values; few individuals in the middle range / 30–35; **[2]**

(c) disruptive; **[1]**

(d) the two populations would move further apart; speciation; **[2]**

(e) may get behavioural isolation / incompatibility; failure to interbreed; **[2]**

13.17 snails predated by birds; banding patterns affect visibility; some snails better camouflaged against their background; there is also climatic selection; dark-coloured shells are better at absorbing heat than light-coloured shells; selection pressures vary through year; no one pattern more successful than the others; **[4]**

13.18 **(a)** A and B have a heavy, short beak; D has a narrow / fine beak / curved beak; C has an intermediate beak, neither as heavy as A and B or as fine as D; **[2]**

(b) adapted to food; finer beaks suited to taking nectar from flower / feeding on fruit; heavy beaks to crack seeds / crush beetles; **[2]**

(c) more niches; wider range of habitats; lack of competition; **[2]**

13.19 selected parent birds with particular / desired characteristics; selected from the offspring those with features similar to that desired and carried out further crosses; repeated for many generations; got gradual divergence from original parents; **[2]**

13.20 **(a)** 56%; 20%; **[2]**

(b) melanic forms camouflaged against darker

bark of trees in industrial area; non-melanic forms more visible; differential predation; **[2]**

13.21 (using the H–W equation $p^2 + 2pq + q^2 = 1$)
$q^2 = 1$ in $2000 = 0.005$;
so $q = 0.0224$ = frequency of recessive allele;
$p + q = 1$
$p = 1 - 0.0224 = 0.977$;
$2pq = 2 \times 0.977 \times 0.0224$
$= 0.044$ or 4.4%; **[4]**

14 Respiration

14.1 **(a)** decarboxylation; decarboxylase; **[2]**
(b) dehydrogenation; dehydrogenase; **[2]**

14.2 a molecule that carries/transfers hydrogen / electrons / carries hydrogen/electrons in reduced form; NAD / FAD; **[2]**

14.3 lactic acid (lactate); **[1]**

14.4 facultative can respire with or without oxygen; obligate cannot survive in the presence of oxygen; **[2]**

14.5 aerobic approximately 36 ATP; anaerobic 2 ATP; (when glucose respired) **[2]**

14.6 **(a)** reduction reaction involves the addition of hydrogen / electron; oxidation reaction involves the removal of hydrogen / electron; **[2]**
(b) reaction in which one compound is reduced and another is oxidised; **[1]**

14.7 **(a)** to absorb the carbon dioxide; **[1]**
(b) mark the level of the manometer fluid in the manometer at the start; record the changes in the level of the fluid at regular time intervals; **[2]**
(c) same temperature throughout; same seeds used in all experiments; same mass of seeds in each tube; same mass of soda lime in each tube; repeat the experiment to calculate a mean; check for leaks; ensure that the level of the liquid in the control apparatus does not move; **[3]**
(d) repeat experiment with one tube containing soda lime, the other containing the same mass of an inert substance; same mass of seeds in each tube; tube with soda lime measures volume of oxygen taken up, second tube measures net gas exchange; the volume of carbon dioxide released is the difference between the two readings; **[3]**
(e) that the substrate is carbohydrate; **[1]** (an RQ of 1 is obtained when an equal volume of oxygen is taken up to carbon dioxide produced, if anaerobic respiration is taking place the RQ value will be larger than 1)

14.8 **(a)**

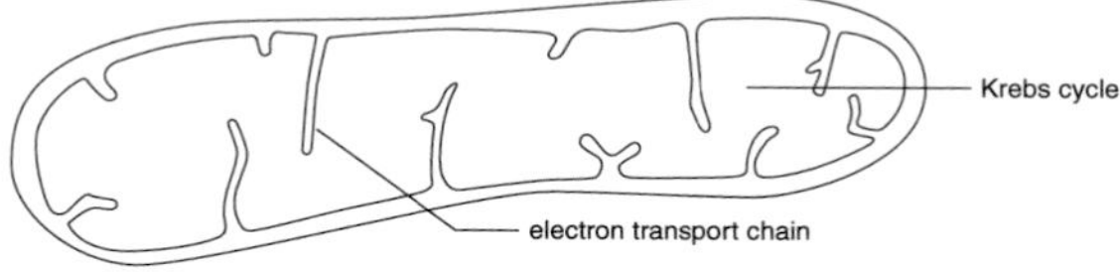

[2]

(b) increase the surface area for the attachment of enzymes / electron transport chain; **[1]**
(c) between 6 and 8 μm; **[1]**

14.9 **(a)** 6 in glucose; 3 in glyceraldehyde 3-phosphate; 3 in pyruvate; **[3]**
(b) 2 ATP; **[1]**
(c) to activate the glucose / to make the glucose more reactive / prevents the glucose from leaving the cell since the membrane is impermeable to glucose-6-phosphate; **[1]**
(d) NAD; NAD^+ picks up H^+ and 2 electrons to form reduced NAD / NADH (may also be written as NADH + H); reduced NAD carries hydrogen and electrons into mitochondrion; **[2]**
(e) cytoplasm; **[1]**
(f) **(i)** taken into mitochondrion for Krebs;
(ii) stays in cytoplasm and undergoes anaerobic respiration; **[2]**

14.10 **(a)** matrix; **[1]**
(b) to break down the acetyl / 2C chain from link reaction; to form carbon dioxide; ATP; reduced NAD / NADH; reduced FAD / $FADH_2$; **[2]**
(c) carbon dioxide is exhaled / ATP used in cellular processes / reduced NAD enters electron transport chain / $FADH_2$ enters electron transport chain; **[4]**

14.11 to generate ATP; re-oxidation of the reduced electron/hydrogen carriers; to oxidise hydrogen to form water; on the inner membrane; **[3]**

14.12 formation of ATP from ADP and P; by a process of oxidation / oxidation of hydrogen and electron carriers; forming water; **[2]**

14.13 **(a)** to allow the yeasts to multiply; **[1]**
(b) carbon dioxide; **[1]**
(c) the ethanol produced by fermentation inhibits growth and cell division; inhibition increases with temperature; low temperatures allow the cells to divide; so there are more of them to ferment the sugars; higher yield of alcohol; **[2]**

(d) high levels of alcohol are toxic to the yeast; yeast die / killed; **[2]**

14.13 (a)

Temperature of water bath / °C	*Time taken for solution to turn pink / minutes*	*Rate of reaction (1 / time)*
0	0	0.000
20	50	0.020
25	36	0.027
30	25	0.040
35	19	0.052
40	13	0.076
45	21	0.047
50	0 (no change)	0.000

(b)

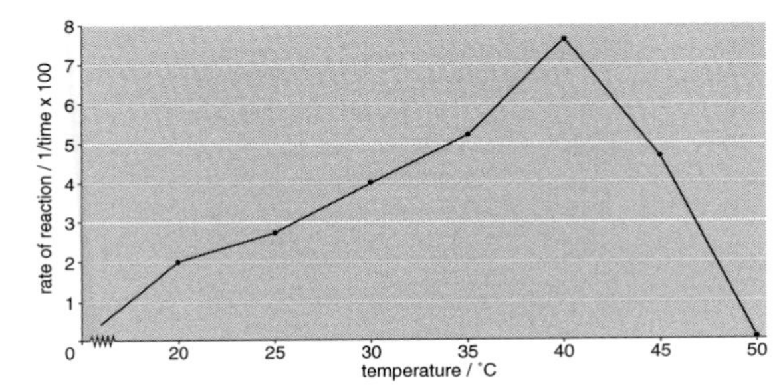

(Note: the rate of reaction values were multiplied by 100 to give easier points to plot.)

Marks are awarded for axes that are suitably labelled; sensible scale so that the graph occupies a large proportion of the graph paper; points plotted accurately; points joined by a neat straight line, ideally drawn in pencil; **[4]**

(c) between 20 and 40 °C as the temperature increases, the rate increases up to a maximum; linear relationship; **[2]**

(d) 40 °C; **[1]**

(e) (Q_{10} = rate at 30 °C ÷ rate at 20 °C)
0.04 ÷ 0.02; = 2 / has doubled; **[2]**

(f) pH constant throughout; same volume of yeast; same volume of TTC; solutions equilibrated before they were mixed; repetitions at each temperature; **[2]**

(g) to remove hydrogen; from the carbon compounds; in Krebs cycle; **[2]**

15 Photosynthesis

15.1 palisade mesophyll; spongy mesophyll; guard cell; **[2]**

15.2 large surface area to absorb light; thin to maximise diffusion because diffusion paths are short; veins support the leaf / network of veins to bring in water and transport away sucrose; mesophyll cells packed with chloroplasts; transparent upper epidermis; palisade cells form single layer beneath epidermis / packed with chloroplasts; thin cell walls / thin layer of cytoplasm to aid diffusion; chloroplasts undergo cyclosis; air spaces to aid diffusion / stomata to allow gas exchange; **[3]**

15.3 (a) A – granum / thylakoids; B – double membrane / chloroplast envelope; C – stroma; **[3]**

(b) light dependent on A / granum; light independent in C / stroma; **[2]**

(c) Length of chloroplast = 30 mm / 3cm
30 ÷ 10 000; = 0.003 mm / 3 μm;
(Method – measure length of chloroplast in mm, divide by the magnification of 10 000 to give the actual value. Check your units. Remember these units are related by a factor of 1000
1 metre = 1000 mm = 1000,000 μm = 1000,000,000 nm)

15.4 both have double plasma membranes; both have large surface area of internal membranes; chloroplast much larger than mitochondrion; mitochondrion has inner folded membrane, chloroplast has unfolded inner membrane; chloroplast contains chlorophyll; chloroplast contains starch grains; both have circular lengths of DNA / ribosomes; inner membrane of mitochondrion has stalked particles attached, no similar structure on chloroplast; **[4]**

15.5 (a) place a bulb at a set distance from the tube containing the pond weed; measure the volume of gas / length of bubble; produced over set period of time; repeat with lamp at three different distances; **[3]**

(b) equilibrate prior to investigation; darken room so no other light source affecting results; no air bubbles in tubing; pinch of sodium hydrogen carbonate to ensure carbon dioxide does not become limiting; repeat investigations to get a mean value; keep temperature constant; **[3]**

(c) (i) ensure no additional carbon dioxide source

in water at start of experiment; carry out experiment at base level, add increasing quantities of hydrogen carbonate to the solution surrounding the pond weed; for each quantity measure the volume of gas evolved;
(ii) use filters / sheets of coloured gelatine across the light source; to ensure only certain wavelengths of light reach the pond weed; **[4]**

15.6 A – water; B – oxygen; C – carbon dioxide; D – sugar/starch; **[4]**

15.7 **(a) (i)** the quantity of light absorbed; by pigments; at differing wavelengths; **[2]**
(ii) the rate of photosynthesis; at different wavelengths of light; **[2]**
(iii) the similarity between the two spectra indicate that the pigments that absorb light; are responsible for photosynthesis; **[2]**
(b) between 420 and 450 nm and at 675 / 680 nm; **[1]**
(c) peaks at 475 compared with 430 nm / chlorophyll *b* peak is higher / 30% higher than chlorophyll *a* peak; peaks at 650 compared with 670 nm; chlorophyll *b* peak is lower / 30% lower than chlorophyll *a* peak; **[2]**
(d) the different pigments absorb different wavelengths of light; more efficient absorption of light; some photosynthesis possible in all light conditions; **[2]**

15.8 **(a)** A – photosystem 2 / P690;
B – photosystem 1 / P700; **[2]**
(b) photolysis / splitting of water; **[1]**
(c) oxygen; NADPH; ATP; **[3]**

15.9 some of the light is reflected by the surface of the water; absorbed by the water; some light converted to heat; not all of the wavelengths of light are absorbed by the pond weeds; **[2]**

15.10 **(a)** with distilled water, rate increases steadily / linear, doubled between 100 and 200; at higher concentrations of sodium hydrogen carbonate the rate of increase is much steeper; at highest concentration, the rate at 200 is four times that of distilled water; **[2]**
(b) a factor is limiting; the rate of photosynthesis cannot go any faster; **[2]**

15.11 **(a)** A – RUBP (ribulose bisphosphate);
B – glyceraldehyde-3-phosphate / triose phosphate; **[2]**
(b) carboxylation; **[1]**
(c) ATP phosphorylates GP (glycerate-3-phosphate); GP reduced by NADPH; **[2]**

15.12 **(a)** the point at which the rate of photosynthesis equals the rate of respiration; there is no net loss or gain of any gas; **[2]**
(b) shade leaves photosynthesise at a greater rate than sun leaves; in a low light environment; **[2]**
(c) shade leaves have a greater uptake initially; then levels off; sun leaves have slower uptake of carbon dioxide earlier in the day; by midday the rate overtakes that of the shade leaves; later in day, the rate of uptake for sun leaves is double that of shade leaves; **[2]**
(d) sun leaves are thicker; have thicker cuticle / thicker epidermis; smaller surface area; thicker palisade layer / more palisade cells; thicker spongy mesophyll layer; **[2]**
(e) the curve would reach a higher maximum rate; as the carbon dioxide would not be limiting; **[2]**

16 Ecology

16.1 The missing words are: *environment*; *biotic*; *temperature* / *water availability* / *salinity* / *pH* / *wind* / *oxygen availability*; *habitat*; **[5]**

16.2 community is all the organisms / plants and animals; living in the same habitat;
population is the number of individuals of the same species; living in the same habitat; **[4]**

16.3 **(a)** diatoms; **[1]**
(b) primary consumer – krill; secondary consumer – blue whale / Adelie penguin / crabeater seal / small fish and squid; tertiary consumer – killer whale / Ross seal / leopard seal / Weddell seal / large fish / emperor penguin; **[3]**
(c) diatoms → krill → small fish and squid → emperor penguin → skua;; **[2]**
or diatoms → krill → crabeater seal → Ross seal → killer whale; **[2]**
(d) food chain only shows one aspect of the feeding relationship; food webs show all the feeding relationships / all the food sources eaten by an animal; **[2]**
(e) decline in large fish / emperor penguins / weddell seals; top consumer, leopard seal would hunt more adelie penguins / crabeater seals; increase in krill; lead to more blue whales / crabeaters / Adelie penguins; **[2]**
(f) high productivity of primary producers / plant plankton eaten whole by consumers, less

waste / leads to more energy available to higher trophic levels; **[1]**

16.4 **(a)** kJ m^{-2} y^{-1}; **[1]**

(b) $470 - 40 = 430$, $40 \div 430 \times 100$; $= 9.3\%$; **[2]**

(c) energy losses from egestion/excretion; not all of an organism is eaten by a consumer; losses due to respiration / metabolism; death of an organism before it is eaten; **[2]**

(d) indicate the quantity of energy locked up in the organisms; biomass does not give any indication of energy; fresh biomass contains a variable water content; biomass values at one point in time; **[2]**

16.5 take random samples; use a quadrat (specify size such as $\frac{1}{4}$ m^2); remove all of the plants and animals from the quadrat area; wash roots of plants to remove soil; separate all the organisms into trophic levels; weigh each of the trophic levels to obtain fresh mass; determine the mean for all the samples; determine the fresh biomass per metre squared; present data as a pyramid; **[6]**

16.6 sampling for biomass only takes place at one point in time; planktonic animals and plants have life span of just a few days; the reproductive rate of the plant plankton is greater than the animal plankton; this is sufficient to support a much greater biomass of animal plankton; **[2]**

16.7 gross is total quantity of energy captured by plant in photosynthesis / rate at which photosynthetic products build up;
net is the quantity of energy available to the primary consumers / net gain of dry mass stored in plant after respiration; net = gross productivity − losses due to respiration; **[2]**

16.8 **(a)** density dependent factors – these factors act on the population; effect proportional to the population size; example – predation / food supply / disease;
density independent factors – factors unaffected by the size of the population; example – climate / flooding; **[4]**

(b) inter-specific is competition between different species; intra-specific competition occurs between individuals of the same species; **[2]**

16.9 **(a)** slow increase over first 4 days; exponential increase from days 5 to 11; levels off at day 12 onwards; fluctuates for remaining 10 days; **[3]**

(b) 400 per cm^3; **[1]**

(c) food supply; space; disease; build up of wastes in water; **[3]**

16.10 **(a)** few doves of breeding age; population scattered; limited the rate of increase; **[2]**

(b) numbers of breeding doves increased; doves exploiting new habitats; not experiencing any competition; **[2]**

(c) reached the carry capacity; no further new habitats to colonise; competition from other species; competition between doves; environmental resistance reducing further increases in population size; **[2]**

16.11 **(a)** plentiful food; availability of nest sites; lack of competition; lack of predation; **[2]**

(b) drought; extreme high/low temperatures; flooding; hurricane or strong force winds; **[2]**

(c) there would be a lack of breeding adults of certain ages; leading to fewer offspring; less stable population; more vulnerable to environmental changes; **[2]**

16.12 **(a)**

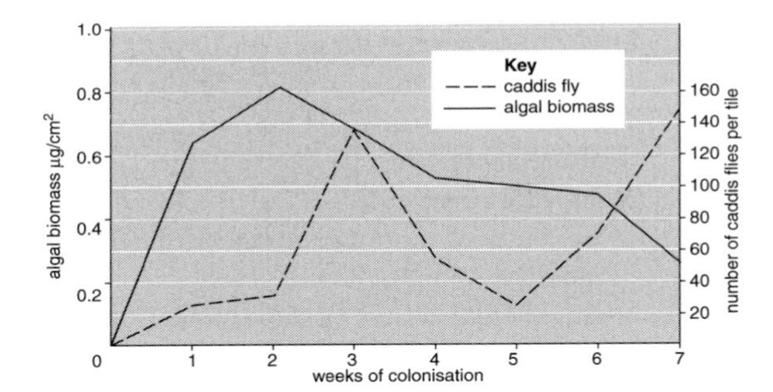

Marks awarded for correct axes with labels; suitable scale so both graphs occupy much of the graph paper; points plotted carefully; points joined up by straight line; key to identify the different curves; **[4]**

(b) both increased initially; peak of caddis fly one week after that of algae; both decreased after reaching peak; caddis flies recovered after week five, but algae continued to decrease; **[3]**

(c) caddis flies are feeding on algae; increase in algae provides more food for caddis fly; more caddis flies move onto tiles; once algae over-grazed, both numbers decrease; algae recover once caddis fly numbers decrease; **[2]**

16.13 **(a)** gradual change in the composition of plant and animal communities in an area; after disturbance or the creation of a new substrate; through a number of stages or seres; increase in biomass / species diversity; leading to a stable / climax community; **[2]**

(b) stable community; often with highest biodiversity; in equilibrium with climate; **[2]**

(c) the range / number of different plants and animals; sum of all the different species; **[2]**

16.14 (a) colonisation by weed/ ruderal species; spread by seeds carried by wind/animals; colonisation by shrubs and small trees/pines/birch; shrubs start to cast a shade over ground; trees form an open canopy over ground; prevents growth of seedlings of many plants/pines; survival of seedlings of plants that tolerate shade / survival of seedlings of deciduous trees; deciduous trees become dominant trees within 150/200 years / climax vegetation; increase in biodiversity; **[3]**

(b) arrival of small shrubs provided greater range of food/nuts/berries/; roosting / nesting sites; cover from predators; varied habitat with open grasses with shrubs favours grassland and woodland bird species; **[2]**

(c) succession reached the final stage / climax vegetation; no further change in plant community; competition for resources; no further niches; **[2]**

(d) make clearings / rides through woodland; coppice trees; more light to reach ground; allow growth of sun-loving plants; greater range of food plants; for different bird species; sites for ground nesting birds; increase in number of niches; **[2]**

(e) interference from humans / disturbance by fire / grazing animals; leads to a different climax vegetation; grazing on grassland prevents climax woodland developing / fires prevent pines colonising heathland; **[3]**

16.15 (a) line transect is the laying of a tape; across an area of vegetation/habitat; species that touch the tape are recorded; either all way along tape or at intervals;
belt transect is a strip across an area; samples taken either continuously along transect or at intervals; **[2]**

(b) frame quadrat is a square frame; enclosing a specific area; for example $\frac{1}{4}$ m^2; count the number of species within the quadrat, or the cover of the species;
point – series of needles; lowered to the ground; record the number of times each species of plant is touched by point; **[2]**

(c) random sampling uses randomly-generated numbers from table or calculator; as co-ordinates; to determine the sample points in a given area;
systematic sampling involves the taking of samples at regular intervals; across the whole sample area; **[2]**

16.16 (a) $30 \div 90 = 100 \div x$;
$x = 90 \times 100 \div 30$; $= 300$; **[3]**

(b) method of marking does not harm/damage the animal; markings do not come off the animals; markings do not make the animals more obvious to predators / change their behaviour; sufficient time elapses between first and second sampling to ensure mixing of the individuals in the population; **[3]**

16.17 lay a transect down the slope; ensure all vegetation zones included; place quadrats at regular intervals along transect; record presence of species; record abundance of species by determining percentage cover / use of domain scale; plot results on graph /kite diagram; **[5]**

16.18 (a) 1 – photosynthesis; 2 – respiration; 3 – respiration; **[3]**

(b) locked up in wood; fossil fuels; sediments in sea; **[2]**

(c) burning of forests / trees; burning fossil fuels; **[2]**

16.19 (a) inert / non reactive; **[1]**

(b) nitrogen-fixing bacteria / *Azotobacter* / *Nostoc* / cyanobacteria; **[1]**

(c) bacteria and fungi; **[1]**

(d) nitrification; **[1]**

(e) 3 – *Nitrosomonas*; 4 – *Nitrobacter*; **[2]**

(f) *Pseudomonas* / *Thiobacillus* / *Clostridium*; anaerobic; waterlogged; **[3]**

(g) nitrogen fixation; convert nitrogen gas into ammonium ions; used by the plant; mutualistic / symbiotic relationship; **[2]**

(h) from soil water; by diffusion; into root hair; **[2]**

(i) proteins / amino acids; DNA / RNA; NADP / NAD; **[2]**

(j) by consuming plants / or other animals; **[1]**

(k) electrical discharges fix nitrogen and hydrogen; to form nitrates in rainwater; **[2]**

16.20 energy from the sun; plants photosynthesise; energy locked up in carbohydrates; plants use the energy for growth/reproduction; gross primary production; animals consume plants; net primary production is gross primary production – respiration; not all of plant eaten by consumer / death of plant causes losses; only 10% of energy transferred to next trophic level; primary consumer eaten by secondary consumer; energy losses from respiration; energy losses from excretion / egestion; role of decomposers / bacteria and fungi; some energy is locked up in fossil fuels; **[8]**

17 Environmental issues

17.1 contamination of the environment with harmful substances / damage of ecological systems; raw sewage / sulphur dioxide / spilt oil / toxic waste / DDT / nitrogen oxide; **[2]**

17.2 protection or preservation of natural resources and environment; management of habitats / maintenance of biodiversity; protecting the environment from human activities; creating new habitats; **[3]**

17.3 **(a)** fuels derived from biomass that was once living / created by the conversion of plant / animal matter over millions of years; contains carbon; stored energy source; examples include oil / gas / coal; **[2]**

(b) increase in number of cars fuelled by petrol; more industry; more people burning fuel in homes / for cooking and heating; **[2]**

(c) sugar cane harvested and milled; produces cane juice; fermentation of juice produces ethanol; benefits – clean burning fuel / no sulphur, produces less pollution than similar quantity of petrol; can be produced close to where it will be used / less transportation; renewable source; carbon dioxide neutral / same amount of carbon dioxide released as was taken up from the atmosphere, does not contribute to global warming; **[4]**

17.4 **(a)** fertiliser run off; raw sewage; treated effluent from sewage treatment works **[2]**

(b) nitrate is soluble; dissolves in water; leached from soil; **[2]**

(c) fertilisers applied during summer months / when crops undergoing growth; rainfall not constant during year / occasional floods; **[1]**

(d) increased use of fertilisers on soil / more intensive farming methods; shift to inorganic fertilisers rather than organic fertilisers such as farmyard manure; organic fertilisers release nitrates more slowly; more people so more sewage; **[2]**

(e) increased growth of algae; algal bloom; algae die; increase in bacteria feeding on dead cells; bacteria respire, use up oxygen in water; high biochemical oxygen demand / oxygen levels fall; active animals such as fish first affected / death of fish and other large animals; disrupted food chain; loss of biodiversity / effect on food web; **[4]**

17.5 **(a)** absorbs ultraviolet light/radiation; ultraviolet light can cause mutations / skin cancer / blindness / cataracts / decrease crop growth; help to regulate atmospheric temperature; **[2]**

(b) CFC (chlorofluorocarbon); aerosol propellant / refrigerant / cleaning circuit boards / expanded polystyrene;
HCFC (hydrochlorofluorocarbon); replacement for CFC;
methyl bromide; pesticide;
halons; fire extinguisher;
tetrachloroethane; solvent in paints / medicines;
methyl chloroform; cleaning solvent / dry cleaning; **[4]**

(c) banning of the production of CFCs / signing of the Montreal Protocol; developing alternatives to CFCs which do not damage ozone; providing aid to developing world to move to CFC free equipment; preventing the trade of CFCs on the black market; **[3]**

17.6 **(a)** unpolluted water tends to have higher oxygen content than polluted water / low biochemical oxygen demand (BOD); high biochemical oxygen demand caused by organic matter / bacteria/ microorganisms in the water; assess by taking sample and determining oxygen use over 5 days; the higher the BOD the more microorganisms / organic material in the water; compare values to samples taken elsewhere along river or at different times of year; **[3]**

(b) some species are very sensitive to pollution; presence or absence of a species can indicate the water condition; sample the animal life in a body of water, identify and check on chart to see the level of pollution that the animal will tolerate; can determine pollution level of the water body; **[3]**

(c) the numerical composition of the habitat; polluted water tends to have a low species diversity/ unpolluted water has high species diversity; determine diversity by collecting sample and recording the number of different species present; calculate index (for example using Margalef index = $S - 1/\log N$ where S is number of different species, N is number of individuals of each species); compare to diversity value obtained on other sites; **[3]**

17.7 **(a)** woodland divided into sectors/areas; each year one area coppiced; majority of standard / large trees cut to ground level; grow back as bushes / multi-stemmed trees; coppiced again 10–20 years later; **[3]**

(b) increases biodiversity; more open nature of coppiced sector allows more light to reach ground; increase in ground plants; bushy stage provides many nest sites for birds; **[2]**

(c) charcoal; fencing; fuel; **[2]**

17.8 **(a)** smallest is Philippines – 3.2%; highest is Venezuela – 71.4%; **[2]** (Note: calculate the percentage loss by dividing primary forest remaining by the original forest cover and multiplying by 100)

(b) demand for more land for housing / roads / industry; clearance for timber / logging; burning forests to clear for farming; mining / quarrying; **[3]**

(c) soil erosion; local flooding; change in local weather patterns; loss of biodiversity; **[2]**

(d) burning of forests releases carbon dioxide into atmosphere; fewer trees to photosynthesise; net result increased carbon dioxide leading to global warming; affects global weather patterns; cleared land surface more reflective, less green canopy to absorb heat; affects air movements / wind patterns; **[2]**

17.9 **(a)** degradation of land; in arid regions of the world; conversion of productive land into desert-like conditions; **[2]**

(b) over-cultivation / over grazing of the land; loss of vegetation cover / increased water run off; loss of soil / reduced organic content of soils; soil erosion;
deforestation / collection of fuel wood; leads to loss of vegetation cover and increased water runoff / loss of soil / reduced organic content of soils; soil erosion;
growing of cash crops on marginal land; crops more susceptible to drought / leads to loss of vegetation cover;
inadequate irrigation; rise in water levels / salts rise to surface / salinisation / soil unsuited to plant growth; **[4]**

(c) re-establishing vegetation cover; tree planting; improved methods of irrigation; mulching crops to retain moisture; establishment of wind shelters to reduce soil erosion; reducing size of grazing herds; **[3]**

17.10 **(a)** sulphur dioxide; nitrogen/nitrous oxide; **[2]**

(b) industry / power stations; vehicle exhausts; **[2]**

(c) **(i)** damage to waterproofing of cuticle on leaf/needles; increased entry of pollutants and pathogens into leaf; chlorosis of leaves / leaves die / die back; reduction in growth rate; trees more prone to frost damage / disease / drought; loss of nutrients from soil leads to reduced nutrient uptake by tree; tree roots damaged leads to reduction in nutrient uptake; **[3]**

(ii) leaching of heavy metal ions / aluminium into water; fall in pH in lake water; acidic water damages fish eggs; aluminium causes fish to produce mucus on the gills / gas exchange reduced; loss of plankton; disruption of food chain/web; loss of biodiversity; **[3]**

(d) other European countries / UK exporting acid rain; acid rain carried on prevailing winds to other parts of Europe; taller chimneys leads to more effective dispersal of acid rain away from source; rain falls on highland of Scandinavia; **[2]**

(e) use of filters in power station chimneys / flue gas desulphurisation; use of low sulphur coal / use of natural gas; increased use of renewable energy; legislation; **[2]**

17.11 **(a)** build up of persistent / non biodegradable pesticides; taken up by consumers in food chain; stored/retained in bodies of animals; passed from consumer to consumer along food chain; higher consumers receive larger doses than consumers lower down the food chain; **[2]**

(b) remains in the environment for a long time / breaks down slowly; **[1]**

(c) pesticide accumulates in bodies of animals in the food chain; especially fatty tissue; top predator eats more food and gets large dose of the pesticide; **[2]**

(d) only affect one pest or group of pests; other organisms unharmed by the pesticide; **[2]**

17.12 **(a)** they are sensitive to sulphur dioxide / air pollution / acid rain; their presence / absence indicates the level of pollution; **[2]**

(b) at 4.5 km from the city centre lichen cover starts to increase; linear / steady increase up to 15 km; then cover levels out; **[2]**

(c) city produces air pollution / sulphur dioxide /

acid rain that damages lichens; city pollution affects an area of approximately 4.5 km diameter; cleaner air / less sulphur dioxide / acid rain further away from city; **[2]**

17.13 **(a)** catch decreases steadily to 1973; even steeper fall from 1974 to 1978; due to over-fishing; catch remains low until 1982; fisheries closed to allow stocks to recover; steep rise / exponential rise in catch from 1982; catch in 1987 exceeds that of 1970; **[2]**

(b) quotas; limiting size of fishing fleet; close seasons; exclusion / protected /conservation areas; protecting spawning / nursery areas; regulating net size; stopping the capture of small / juvenile fish; **[3]**

18 Homeostasis

18.1 maintenance of the internal environment; involvement of co-ordination and control mechanisms; use of negative feedback mechanisms; changes detected by receptors; example – osmoregulation / excretion / blood glucose levels / temperature; **[3]**

18.2 detection of increase/decrease of a substance away from the norm; response brings levels back to the norm; positive feedback takes levels further away from the norm; an increase/decrease causes changes which bring about further increases/decreases; **[3]**

18.3 **(a)** excretion – the removal of waste products; which would become toxic if allowed to remain in the body; osmoregulation – control of the water balance in the body; **[3]**

(b) lungs – carbon dioxide; kidneys – urea/excess salts; liver – breakdown products of red blood cells / biliverdin / bilirubin; **[3]**

18.4 A – renal artery; B – renal vein; C – cortex; D – medulla; E – ureter; F – pelvis; **[6]**

18.5 liver; **[1]** by deamination; amino group removed from amino acid; produces ammonia; ammonia converted to urea; in ornithine cycle; **[2]**

18.6 specialised nerve cells; sensitive to concentration of plasma / water potential of the blood; in the hypothalamus; **[3]**

18.7 receives impulses from hypothalamus; causes pituitary to release ADH (anti-diuretic hormone) into blood; travels to kidney; where it increases permeability of the second convoluted tubule and collecting duct; **[2]**

18.8 temperature at the centre of the body / deep in the body; kept within narrow range of just a few degrees; 36–38 °C; **[2]**

18.9 conduction; convection; radiation; **[3]**

18.10 The missing words or numbers are: 37; skin; nerve impulses; increase; evaporation; vasodilation; increases; radiation; **[8]**

18.11 **(a)** A – glomerulus; B – collecting duct; C – tubule;

(b) diameter on photo = 22 mm
actual diameter = measured diameter / magnification = 22 / 300;
= 22 mm / 300 = 0.07 mm or 70 μm; **[2]**

18.12 vasoconstriction / arteriovenous shunt opens; to reduce blood flow through capillaries near the surface of the skin; hairs raised / pili erector muscle contracts; air trapped around skin to reduce heat loss; shivering; increase in metabolic rate; to generate heat energy; **[3]**

18.13 hypothermia; **[1]**
(Note the spelling – many candidates spell this word incorrectly in examinations)

18.14 The missing words are: beta; Langerhans; insulin; glycogen; muscles; glucagon; alpha; **[7]**

18.15 fish live in aquatic habitat; their nitrogenous waste product is ammonia; is toxic; needs to be diluted in large quantities of water; diffuses from blood to water in the gills; insects terrestrial and water less freely available; uric acid is a solid; can be passed out in faeces; without using water; **[3]**

18.16 fall in blood glucose levels below normal; **[1]**

18.17 **(a)** both have a similar shape / increase followed by decrease; all the values of blood glucose for curve A are higher than those for curve B at corresponding times; values for curve B return to the starting value, curve A remains higher at end; curve A increases more steeply than curve B; **[3]**

(b) A; rises more steeply than B; does not return to original value; slow decrease in glucose levels; **[3]**

(c) failure to synthesise insulin; due to lack of insulin secreting cells in pancreas / destruction of beta cells in islet of Langerhans due to viral infection; inability to produce sufficient insulin, as a result of ageing / failure to control blood glucose / failure of insulin secreting cells to respond to glucose levels; **[2]**

(d) injection of insulin; control of the carbohydrates in the diet (mild cases); **[2]**

18.18 purer; fewer side effects; cheaper; large quantities produced in short time; **[3]**

18.19 **(a)** ectothermy – external heat source; body temperature changes / fluctuates with environmental temperature; controlled by behaviour;
endothermy – body temperature maintained within narrow limits; independent of environmental temperatures; heat source is metabolism; regulation by physiological mechanisms; **[3]**

(b) A; body temperature changes with environmental temperature; not regulated; **[3]**

(c) ectotherms have to live in an environment which can supply them with heat / need to live in a warm place; need to sun themselves each morning to warm up; large ectotherms such as lizards not found in far north and south; many inactive over winter months; **[2]**

18.20 **(a)** squeezing of blood plasma; through pores in the capillary; due to build up of high pressure; **[2]**

(b) narrower diameter; increases pressure inside capillary; forces plasma out of the blood; into the Bowman's capsule; **[2]**

(c) molecules under MW 68 000; glucose / amino acids / ions / water / urea; **[2]**

18.21 filtrate in the descending limb of loop of Henle flowing in opposite direction to filtrate in ascending limb; cells of ascending loop actively pump out chloride ions; sodium ions follow; this makes fluid around loop more concentrated; sodium and chloride ions diffuse into descending limb; filtrate in tubule is most concentrated when it reaches the top of the loop; descending limb permeable to water; water leaves descending limb; water passes into surrounding capillaries / vasa recta; **[4]**

18.22 **(a)** to produce a concentrated filtrate / to absorb as much water as possible; **[1]**

(b) water released by metabolism; when glucose is respired; **[2]**

(c) active at night when cooler; stay in burrow during day; burrows are cool and sheltered so less water loss; **[2]**

18.23 **(a)** hypothalamus; **[1]**
(Note: ADH is produced in the hypothalamus then secreted into and stored in pituitary gland, from where it is released)

(b) when blood becomes more concentrated / water potential of blood more negative / after sweating / after a salty meal; **[1]**

(c) increases the permeability; of the second / distal convoluted tubule and collecting duct; more water moves into surrounding tissues / blood vessels; **[2]**

18.24 **(a)** 99.2; 99.9; 50; 99.4; 99.2; 86.1; % **[6]**
(Note: divide the quantity reabsorbed by the quantity filtered and multiply by 100, e.g. 178.5/180 × 100%)

(b) required by body for respiration; too valuable to be excreted; **[2]**

(c) diabetes; **[1]**

(d) concentration of urea higher in filtrate; so diffuses into blood / diffuses down concentration gradient; **[2]**

18.25 **(a)** brush border / microvilli / highly folded membrane; folded basement membrane; to increase surface area; numerous mitochondria; to release energy for active transport; **[2]**

(b) glucose moved by active transport; from filtrate into cell; diffuses through cell; actively moved across basement membrane; diffuses into capillary; movement of solutes out of tubule creates osmotic gradient; water moves by osmosis; **[2]**

18.26

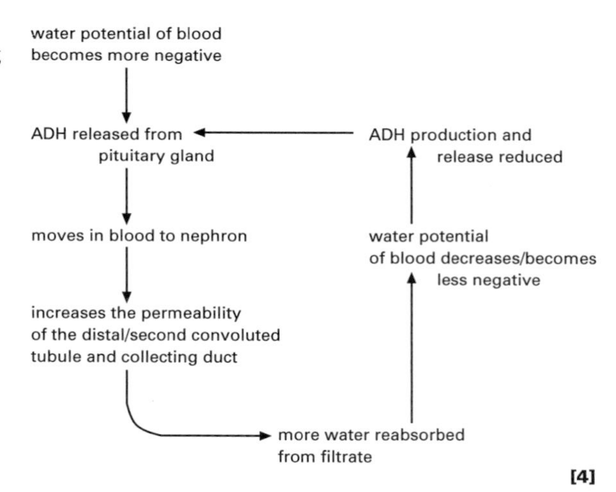
water potential of blood becomes more negative
ADH released from pituitary gland
ADH production and release reduced
moves in blood to nephron
water potential of blood decreases/becomes less negative
increases the permeability of the distal/second convoluted tubule and collecting duct
more water reabsorbed from filtrate

[4]

19 Nervous and chemical co-ordination

19.1 **(a)** chemical messenger; secreted by endocrine gland; controls processes in the body; made of protein/amine/steroid; **[2]**

(b) ADH; hypothalamus / secreted from pituitary; kidney;
insulin; beta cells of islets of Langerhans; liver;

adrenaline; adrenal medulla; widespread effect in body including liver / muscles / heart; follicle stimulating hormone; pituitary gland; ovary; **[3]**

19.2 part of the nervous system consisting of the brain; and spinal cord; **[2]**

19.3 **(a)** presence / change in concentration of a specific substance in the blood; nervous stimulation; a change in concentration or the presence of another hormone; **[2]**

(b) a rise in blood glucose triggering release of insulin / presence of oestrogen triggers release of luteinising hormone / nervous stimulation causing the release of adrenaline from adrenal gland; **[1]**

19.4 the missing words are: hormones; steroid/amine; endocrine glands; bloodstream/system; target; **[5]**

19.5 some hormones bind to receptor on cell surface membrane; cause the release of second messenger inside cell; some hormones diffuse through cell surface membrane; travel to nucleus where they activate a gene; some affect membrane permeability allowing substances to enter cell; **[2]**

19.6 **[5]**

Nervous control	*Chemical control*
information in form of electrical impulse	information as chemical messenger;
response immediate	response slow;
response short-lived	response long-lasting;
transmission fast	transmission slow;
effect localised	effect can be widespread;

19.7 its secretions regulate all the other endocrine glands; either directly or indirectly; **[2]**

19.8 **(a)** A – dendrite; B – nucleus; C – axon; D – myelin sheath; E – node of Ranvier; F – cell body; **[6]**

(b) carry impulses from central nervous system; to effector / muscles; **[2]**

(c) sensory; relay/connector; **[2]**

(d) acts an electrical insulator; speeds up transmission of impulses; **[2]**

19.9 **(a)** A – dorsal root ganglion; B – dorsal root; C – central canal; D – grey matter; E – white matter; **[5]**

(b)

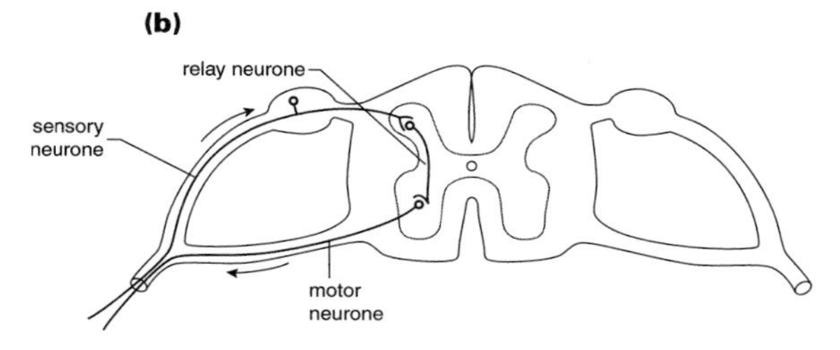

Marking – to gain the full marks, the cell body of the sensory neurone needs to be shown in the dorsal root ganglion, it synapses with a relay neurone inside the grey matter, this relay neurone synapses with a motor neurone, also within the grey matter, the cell body of the motor neurone is within the grey matter **[3]**

(c) shown on figure **[1]**

19.10 **(a)** if there is too much thyroxine, it inhibits/suppresses the action of the hypothalamus and the pituitary; so less thyroxine is produced; if there is too little, the hypothalamus and pituitary are not suppressed; and thyroxine is produced; **[3]**

(b) negative feedback; **[1]**

(c) when it affects two, it gives greater control / sensitivity; **[1]**

(d) in the blood stream; **[1]**

19.11 to control / co-ordinate growth; to control / co-ordinate development; **[2]**

19.12

Plant growth substance	*Site of synthesis*	*Effect in plant*
auxin	apical meristem (shoot tip) / young leaves;	stem elongation / lateral root initiation / apical dominance (inhibits growth in lateral buds) / involved in fruit formation;
gibberellin	throughout shoot / seeds;	stem elongation / bolting / fruit development / seed germination;
cytokinin	meristem / root apex / shoot apex (i.e. places where rapid cell division taking place) / endosperm of seeds;	development of roots / development of lateral buds / cell enlargement / delays senescence;
abscisic acid	most organs of the plant / seeds;	abscission (leaf fall) / seed dormancy / stomatal opening and closing in stress conditions;
ethene	throughout plant / fruits;	ripening of fruits and seeds / abscission / wound healing / breaking dormancy in seeds / sprouting in potato tubers and bulbs;

[10]

19.13 weedkiller; rooting powders for propagation; development of seedless fruits; delays abscission in fruits / fruit drop; **[2]**

19.14 absorption of water by seed results in production of gibberellin; diffuses into aleurone layer; stimulates synthesis of enzymes; to hydrolyse starch / food stores; starch broken down into maltose and glucose; used in respiration; **[3]**

19.15 A – cerebral hemisphere / cerebrum; receiving sensory information / controlling voluntary muscle movement / mental activity / language / speech / memory;
B – medulla oblongata; controls heart beat / cardiac output / control of breathing / peristalsis / coughing / sneezing / swallowing / vasodilation and constriction;
C – cerebellum; co-ordination of skeletal muscles / balance and posture;
D – spinal cord; conduction of impulses to and from brain; **[8]**

19.16

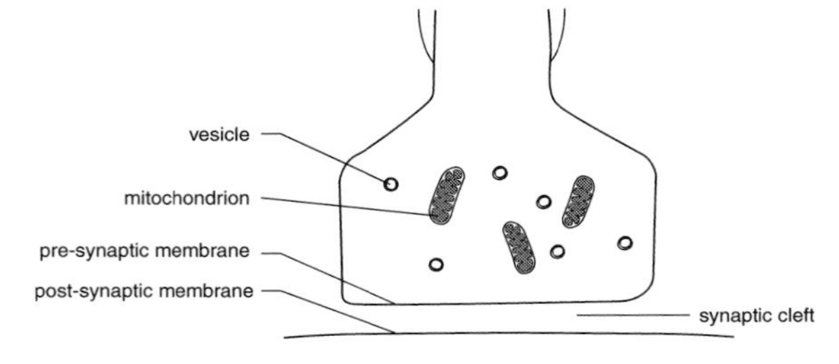

3 marks are given for accuracy and quality of drawing, 5 for correct labels **[8]**

19.17 **(a)** antagonistic is two chemicals/substances working against each other / effects of one substance counteracted by the other; synergistic is working /acting together / effects of one substance enhanced by the other; **[2]**

(b) auxin + gibberellin /
auxin + cytokinin = synergistic;
auxin + abscisic acid /
auxin + ethene = antagonistic; **[2]**

19.18 arrival of impulse opens calcium channels in pre-synaptic membrane so calcium diffuses into synaptic knob; calcium stimulates movement of vesicles toward pre-synaptic membrane; vesicles contain transmitter / acetylcholine; vesicles fuse with membrane; transmitter released into synaptic cleft; transmitter diffuses across cleft and binds to receptors on post-synaptic membrane; ion channels open; post-synaptic potential generated; **[6]**

19.19 **(a)** −70 mV; **[1]** (make sure you include the units)

(b) differences in ion concentration either side of membrane; due to differential permeability of the membrane; membrane permeable to potassium due to protein channels/gates, so potassium accumulates inside axon; no channels for negative charged ions which are trapped inside; low permeability to sodium ions, so they remain outside; **[3]**

(c) permeability to sodium increases / sodium gates open; sodium ions flood into axon; **[2]**

(d) permeability to sodium decreases / sodium gates close; sodium ions pumped out of axon;

increased permeability to potassium; potassium ions flow into axon; **[3]**

(e) period during which the resting potential is restored; **[1]**

(f) the longer the refractory period the fewer the number of impulses; **[1]**

(g) too many potassium ions enter membrane / potassium gates remain open; overshoot the resting potential; **[2]**

19.20 The missing words are: cornea; accommodation; short; fat; ciliary muscles; suspensory ligaments; retina; rods; rhodopsin; cones; fovea; optic; **[12]**

19.21 in bright light, circular muscle contracted and radial muscles relaxed; pupil is small and limits amount of light entering eye; in low light, circular muscles relaxed and radial contracted; pupil dilated / enlarged; more light enters eye; **[3]**

19.22 three cones, each sensitive to a different colour of light; each contains a different form of iodopsin; combinations of cones means that wide range of colours can be detected; **[3]**

19.23 **(a)**

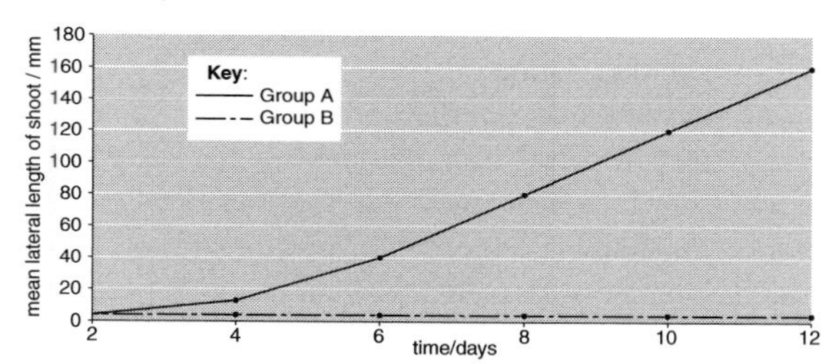

Marks are awarded for correct axes with labels; suitable scale; points plotted accurately; points joined up with straight line; line labelled / key; **[5]**

(b) slow increase in shoot length over first two days; linear / steady increase from day 4; **[2]**

(c) produce auxin; auxin diffuses back down shoot; inhibits development of lateral buds; **[2]**

(d) all conditions the same, including light direction and levels, temperature; accurate measurement of the shoots, measuring from the same point on all plants; experiment repeated to obtain a mean for each treatment; **[2]**

20 Synoptic questions

20.1 **(a)** point mutation (specifically a substitution); **[1]**

(b) single base replaced by another in one codon; results in a code for a different amino acid; changes the sequence of amino acids in the polypeptide chain; **[2]**

(c) affinity means attraction; HbS is less attracted to oxygen than HbA; HbS carries less oxygen; **[2]**

(d) having equal dominance / neither dominant nor recessive; both alleles expressed in heterozygote; **[2]**

(e) no sickle cell or malaria in Southern Africa / much of central North Africa; areas of highest incidence of sickle cell only in areas where malaria occurs; only low incidence of sickle cell in areas where malaria does not occur; **[2]**

(f) no malaria; no advantage to carrying a sickle cell gene; genetic – nobody carrying the gene in the population of South Africa; **[2]**

(g) have small number of sickle-shaped red blood cells; presence of the sickle-shaped red blood cells/ lower oxygen concentration prevents infection by the malarial parasite; heterozygotes more likely to survive than either of the homozygotes; gets resistance without the full blown disease; **[2]**

(h) no advantage to carrying HbS gene so frequency will decrease with each generation; level out at a very low frequency or may eventually disappear completely; **[2]**

20.2 **(a)** energy store; respired during germination; to fuel the early growth of the seedling until it can photosynthesise; **[2]**

(b) translocated; along phloem sieve tubes; from source/leaves to sink; **[2]**

(c) cytoplasm; **[1]**

(d) used in Krebs cycle of respiration; acetyl part joins 4-carbon compound to form a 6-carbon compound; **[2]**

(e) contains –COOH / carboxyl group; general formula of R.COOH where R is hydrogen or $-CH_3$; hydrocarbon chain / chain of carbon and hydrogen atoms; large number of carbon atoms; **[2]**

(f) phospholipids; **[1]** they are polar; hydrophilic heads; hydrophobic tails; form lipid bilayer; allow organic solvents to pass through membrane; **[3]**

(g) renewable source; carbon dioxide neutral – don't contribute to global warming; easier to obtain compared with drilling for oil; are more complex so need less processing at refineries; refining process requires less energy and produces less pollution; vegetable oil is biodegradable / mineral oils have impurities such as sulphur which do not degrade quickly; vegetable oils produce less sulphur dioxide / less pollution when burnt / clean burn fuels; **[3]**

(h) genetic engineering; identify a gene for modification enzyme in another oil crop; remove it using restriction enzymes; insert in plasmid; add marker gene such as gene for antibiotic resistance; insert plasmid in *Agrobacterium tumefaciens*; infect cells of the sunflower with the *Agrobacterium* containing the new gene; select sunflower cells with new gene; grow in culture; **[4]**

20.3 molecular structure of ATP
source of chemical energy, same in all organisms
universal energy currency
produced in respiration / role of mitochondria
phosphorylation of ADP to form ATP
hydrolysis of ATP yields large quantity of energy
ATP continually resynthesised, ATPase
active cells have more mitochondria, for example liver, muscle
ATP needed for:
synthesis of macromolecules / proteins / carbohydrates
movement, muscular contraction
active transport
activates glucose in glycolysis
photosynthesis – produced in light dependent stage / photophosphorylation and used in light independent stage
[15] marks given for scientific content and writing style

20.4 overview of DNA structure
only found in nucleus
semi-conservative replication ensures DNA the same in new cells
genetic code / codons
mRNA / tRNA and protein synthesis
gene – sequence of nucleotides on DNA that codes for a polypeptide
polypeptides form enzymes
expression of genes / dominant and recessive / genotype / phenotype
changes in the genetic code leading to point mutation
examples of point mutation
role of mutations in evolution
chromosomes and DNA
mitosis and meiosis
brief reference to gene technology
[15] marks given for scientific content and writing style

20.5 covers $\frac{2}{3}$ of the planet
life evolved in water
properties of water and how they relate to organisms (solvent, adhesion/cohesion, surface tension, freezing and boiling point, latent heat of vaporisation, etc)
water makes up large proportion of organisms / cells
water as a habitat for aquatic organisms – thermal properties, freezing point
water as a solvent
soil water source of minerals for plants
chemical reactions, hydrolysis, role in metabolic processes
surface tension
transport medium in plants and animals
osmosis
transpiration
raw material for photosynthesis – photolysis
osmoregulation
fertilisation – external and internal
seed dispersal in aquatic plants
support – hydrostatic skeletons in animals, maintaining turgor in plants
lubricant – tears, synovial fluid
cooling – sweat and heat loss
[15] marks given for scientific content and writing style

20.6 carbohydrate definition and examples
chemical formula
mono-, di- and polysaccharides
uses of monosaccharides – respiratory substrate, intermediates in photosynthesis, nucleic acids, NAD, NADP, FAD, synthesis of AMP, ADP and ATP
disaccharides – milk sugar, translocated in sieve tube cells of plant, storage in sugar beet and sugar cane
polysaccharides – starch, cellulose, glycogen
starch – only in plants, brief overview of structure, amylose and amylopectin, insoluble and osmotically inert so can be stored, energy source
cellulose – structure, tensile strength, permeable, plant cell wall, food source for ruminant herbivores which have cellulase enzymes in rumen
glycogen – structure, animal equivalent to starch,

energy store, muscles, liver
other carbohydrates – pectin(cell wall), murein (bacterial cell wall), inulin and mannan (food storage) mucilage and gums (protective roles)
related compounds – chitin in fungi and arthropods (exoskeleton)
glycoprotein and glycolipids – cell recognition
[15] marks given for scientific content and writing style